GETTING TO KNOW SEMICONDUCTORS

GETTING TO KNOW SEMICONDUCTORS

M. E. Levinshtein

G. S. Simin

Published by

World Scientific Publishing Co. Pte. Ltd.
P O Box 128, Farrer Road, Singapore 9128
USA office: Suite 1B, 1060 Main Street, River Edge, NJ 07661
UK office: 73 Lynton Mead, Totteridge, London N20 8DH

Library of Congress Cataloging-in-Publication data is available.

The authors and publisher are grateful to the following publishers for their permission to reproduce the reprinted figures found in these volumes:

American Institute of Physics (*Physics Today*)
Lake Shore Cryotronic, Inc.
McGraw Hill Inc. (*The Strain Gauge Primer, Electronics*)

GETTING TO KNOW SEMICONDUCTORS

ISBN 981-02-0760-3

Printed in Singapore by Utopia Press.

PREFACE

Getting to know semiconductors is very important and useful. Semiconductors, the subject of our book, are found in every home. They also have very good connections in different spheres: industrial, scientific and military which are so wide that they are beyond the imagination of an ordinary man.

But it is common knowledge that making somebody's acquaintance and keeping it up just because it is useful is both bad style and dull.

Semiconductors is a very useful and most interesting subject. So we decided to do our best to assist our readers in getting to know them.

Our friends and colleagues have done a lot to provide a conducive atmosphere for the writing of this book. We are very grateful to all of them.

We are specially obliged to Prof. M. I. Dyakonov (*A. F. Ioffe Institute, Leningrad, Russia*) and Prof. B. I. Shklovskii (*Univ. of Minnesota, Minneapolis, USA*). We are also most thankful to Ms. M. A. Simina and Ms. L. G. Titova for their help, support and constant attention to our work.

We would like to conclude with the wonderful words from the Gospel: "Ask, and it shall be given you; seek, and ye shall find, knock, and it shall be opened unto you (*Matthew 7:7*).

CONTENTS

GETTING TO KNOW SEMICONDUCTORS

INTRODUCTION

> Bodies, able to conduct electricity, and being something between conductors and insulators, are usually called semiconductors.
>
> Ivan Dvigubsky
> *Basic Experimental Physics,* 1826

> The term semiconductors will be used for those conductors, whose resistance is greatly affected by temperature.
>
> I. Köenigsberger, 1914

Nowadays it is perhaps impossible to meet a man who has not heard anything of semiconductors. Almost everyone is sure to know that the core of the most important device of modern civilization – the computer – is nothing but a small semiconductor (silicon) plate, on which hundreds of thousands and sometimes even millions of tiny semiconductor devices – transistors – are placed.

Almost everyone might have suffered from his neighbor's tape recorder or record player switched on at their loudest. And everyone enjoys listening to his favorite music, switching on his own tape recorder or record player as loud as he likes. A modern car is known to have tens of different semiconductor devices. Everyone might have happened to see a photograph of the Sputnik (artificial satellite) with wide "wings", containing the solar batteries to feed the devices on board the Sputnik.

The definition of the word "semiconductor" given in the encyclopedia or in a reference book is similar to that given in the epigraphs to our Introduction. Firstly, the conductivity of these materials is neither very large, nor very small. It is smaller than the conductivity of good metals such as copper, silver, aluminium and iron. But it is much greater than the conductivity of such good dielectrics as glass, wood and paper. Secondly, the conductivity of these substances depends to a very great extent on temperature. To understand why these two, seemingly neither very important nor very interesting properties of semiconductors ensure their brilliant and dashing career, we have to write and you have to read this story. Two particles – the electron and the hole will be the main heroes.

THE MAIN HEROES: ELECTRONS AND HOLES

Oh! What familiar faces we see here!

A. S. Gryboedov

The names of the main heroes of our book are the electron and the hole.

The electron is a hero of not only a "semiconductor novel". Electrons are met in cosmic rays and in the accelerators of elementary particles as well as in metals and in vacuum devices (such as TV kinescopes and tubes).

The activities of a hole are much more limited. More often we come across holes in semiconductors.

The hole is much younger than the electron. The term "electron" was introduced in 1891 by the Irish physicist Jonston Stoney to denote the particle carrying the elementary charge of the negative electricity 1.6×10^{-19} Coulomb.

The term "hole" was introduced in 1933 by the Soviet physicist Jacob I. Frenkel to denote "the particle" able to create a current in semiconductors and carrying an elementary positive charge, equal in absolute value to the charge of the electron.

It is not accidental that when applied to the hole the word "particle" is in inverted commas. Generally speaking, the hole is not a particle.

The hole has its charge and mass. We can read about the concentration of holes in semiconductors, about the trajectory of a hole in the electric or magnetic field, or about the interaction of holes and electrons. But ... as a matter of fact there is no such particle in nature. The notion "hole" merely denotes the absence of an electron where it should be.

In order to understand why an empty place, a hole, is called particle and is ascribed mass and charge, we have to study very thoroughly the field of action of our heroes, the semiconductors.

We will begin with a short digression.

Atoms of any substance are electrically neutral. If we apply an electric field to a volume filled with neutral particles, the electric current will not flow through that volume. Since no charged particles are present, there will not be any electric current, which is in fact the directed motion of charged particles. Thus, the volume, filled with atoms of any substance, is in fact an ideal insulator.

The air is a good example of such an insulator. Every cubic centimeter of air contains 2.7×10^{19} molecules of oxygen (O_2), nitrogen (N_2), vapor (H_2O) and some other gases which we will just ignore. Every atom of oxygen contains 18 positively charged protons and the same number of negatively charged electrons. Every atom of nitrogen contains 7 protons and 7 electrons. It might seem there are more than enough charged particles there, but those particles are bound by powerful electric forces to form electrically neutral atoms and molecules, as a result of which the air is a perfect insulator. High-voltage lines stretch for thousands of kilometers transmitting energy with negligible losses. A well-designed battery can stay in the air for years without being discharged. A volume filled with a gas of neutral atoms of silver, copper, gold or mercury will also be a splendid insulator. But why do those substances, when they are in the solid state, have quite different conductivities? Solids seem to be composed of the same atoms. And still the conductivity of a good conductor, for instance of silver, is about 10^{22} times greater than the conductivity of a good insulator (for instance of glass), which is in fact ten thousand billion billion times!*

Moreover, the atoms of one and the same substance, e.g. of carbon (C), depending on the type of crystal they form, can be either a very good conductor (graphite), or a perfect insulator (diamond).

Metals, Dielectrics and Semiconductors

The last example suggests that whether the solid is a metal or a dielectric depends not so much on the properties of the atoms forming the crystal,

---

*To realize better the difference between the conductivity of good metals and that of good insulators one should note that it is absolutely the same as the difference between the diameter of our Galaxy and 1 cm ($\sim 10^{22}$ times).

as on the types of the bonds of atoms in the crystal lattice of the solid body. Figure 1 demonstrates the crystal lattice of the metal, Fig. 2 – of the dielectric.

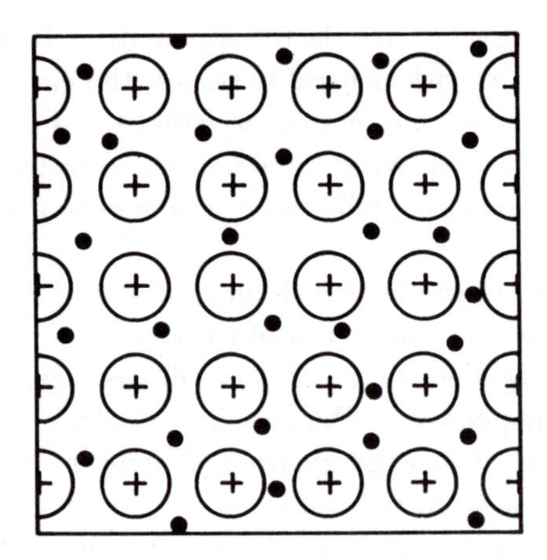

Fig. 1. Schematic diagram of the metal crystal lattice. The regular lattice of positively charged ions is plunged into the "gas" of free electrons, having no tight bonds with separate ions.

Figure 1 illustrates the main properties of metallic crystals. The crystal lattice is formed by positively charged ions, not by neutral atoms. While forming the lattice, each atom loses one valence electron, and these electrons (shown as black circles in Fig. 1) do not belong then to any specific ion of the metal. These electrons are said to be collectivized by the crystal and can move freely under the action of the external electric field. There are $\sim 10^{22}$ free carriers – electrons – in every cubic centimeter of the metal. It is no wonder that metals are perfect conductors.

The scheme of the crystal lattice of silicon, shown in Fig. 2, demonstrates another type of bond in a crystal. Every atom is linked with the neighboring atoms by means of strong electron bonds.

The electron bonds of an arbitrarily-chosen atom 1 with his four neighbors is shown in detail in the center of Fig. 2. Eight electrons rotate round the atom in closed orbits, whose form is rather complicated. Four of the eight electrons have always belonged to atom 1: we know silicon to have four valent electrons on the outermost electronic shell. But what about the other four electrons? Where do they come from? It is seen from Fig. 2 that they come from the neighbors in the crystal lattice. But atom 1 has not appropriated these electrons: they spend some time near atom 1 and then they return to their host and spend the rest of the time with their primordial host. On the other hand, the valence electrons of atom 1 now spend only part of their time with atom 1, and the rest of the time is spent near the neighboring atoms. As a result, on the whole, the negative charge of the electrons round atom 1 is still equivalent to the charge of the four electrons. Now, however, atom 1 is linked to atoms 2, 3, 4 and 5 (and those, in their turn, to their neighbors and so on) by strong electron bonds.

Fig. 2. Schematic diagram for the silicon (Si) crystal lattice. The lines linking the Si atoms represent electron bonds.

The fields, keeping the electrons in their orbits, are very strong. Let us estimate the field acting on the valence electrons on the part of the atom of silicon. The number of silicon in the Periodic Table is 14. That means that the nucleus of the atom of silicon contains fourteen protons. As a whole the atom is neutral: the number of protons equals the number of electrons, while the charge of the proton is equal to the charge of the electron in magnitude but is opposite in sign. As we know, there are four valent electrons on the outer electron shell of the silicon. The other ten electrons are located on the inner shells and partly screen the action of the electric field of the nucleus onto the valent electrons. Thus, the action of the nucleus upon every valence electron is equivalent to the attraction on the part of four protons.

Let us assume that the electric field created by the nucleus and by the electrons of the inner shells acts upon the valence electron as a field of a point charge. We will assume the average distance between the atom and the valence electron to be equal to the distance between the atoms (Fig. 2). Then, from Coulomb's law we can obtain the average value of the intensity of the field keeping the valent electron in its orbit:

$$F \cong \frac{4q}{4\pi \epsilon_0 a_0^2}$$

where q is the electron charge, $\epsilon_0 = 8.85 \times 10^{-12}$ F/m is the permitivity of vacuum, a_0 is the distance between the atoms of the crystal lattice. For the majority of solids a_0 is about a few fractions of the nanometer (1 nm $= 10^{-9}$ m). In silicon $a_0 = 0.54$ nm. Substituting into the equation the known values of q, ϵ_0 and a_0, we obtain $F \cong 2 \times 10^{10}$ V/m.

To imagine the value of the electric field $F \sim 10^{10}$ V/m, we can say that the field with the intensity of $F \approx 10$ V/m can heat an iron nail so much that it will melt down. (Calculate the density of the current in an iron wire when $F = 10$ V/m). And we can mention here that even Zeus the Thunderer has at his disposal electric fields, equal, approximately, to only 3×10^6 V/m. As soon as the electric field between the clouds and the Earth reaches that value, there is thunder and lightning in the atmosphere. But we must emphasize that the fields of the thunderstorms are about 10 000 times weaker than the electric fields which are keeping the electrons in their orbits.

Now we are approaching one of the most essential questions. If we apply an external electric field to a crystal, similar to that shown in Fig. 2, will it give rise to an electric current?

The answer is no, of course not! Such a crystal is in fact an ideal dielectric, since even a strong electric field ($\sim 10^6$ V/m) only slightly deforms the orbits of the electrons, but it is unable to break them. There won't be any free charge carriers in the crystal, so there won't be any electric current either.

We'll note that it isn't simple to create such an ideal lattice which is shown in Fig. 2. For the lattice to be really ideal, it is necessary to exclude any deformation, imperfections or defects which might break the electron bonds between the atoms. Besides, the crystal must be absolutely pure, there should be no impurity there. And, last of all, the crystal must be cooled down to the temperature of the absolute zero, otherwise the temperature fluctuations of the lattice may break the electron bonds and free carriers may appear. All those factors: temperature, impurities, and defects are of great importance and further on we'll discuss them in detail. Here we'll note that at the temperature of the absolute zero, a crystal, similar to that shown in Fig. 2, is in fact an ideal dielectric. So, we have considered two types of crystals. Some of them (Fig. 1) are very good conductors of the electric current. Other crystals, at low temperature, do not conduct any current at all (Fig. 2). Those are dielectrics.

Now it is time to discuss the third type of crystals which is the most interesting for us – the semiconductor crystals.

However... remember that Fig. 2 shows the crystal lattice of the most typical and wide-spread semiconductor – the silicon. And near the absolute zero such a crystal is, as we have already mentioned, an ideal dielectric.

Here we can confess that there is no principal difference between semiconductors and dielectrics.

The model of the crystal, shown in Fig. 2, can be applied to both: to typical dielectrics and semiconductors. The difference between them is purely quantitative. It is determined by the value of energy necessary to break the electron bonds between atoms. The greater the energy needed to make the electron leave its orbit, and thus make it a free electron able to move under the action of the electric field, the more grounds we have to say that we deal with a dielectric.

First let's estimate what energy we must use to make an electron free. We have calculated that the field keeping the electron in its orbit is $F \sim 10^{10}$ V/m. The force acting on the electron $f = qF$. To eliminate the electron from its orbit it is necessary to counteract this force f and to "pull" the electron away from the ion attracting it, so that its distance from the ion should be approximately equal to the lattice constant. Thus, the energy necessary to break the electronic bond and to make the electron free, is $E_g \cong qFa_0$. The index g comes from the word "gap". Substituting the known data of q, F and a_0 into the formula we obtain $E_g \cong 8 \times 10^{-19}$ J $\cong 5$ eV.* The magnitude E_g depends in fact on the structure of the crystal lattice and

*A unit of energy of 1 electronvolt (1 eV) is equal to 1.6×10^{-19} J. The electron acquires the energy of 1 eV after it passes the potential difference of 1 V. When we speak of the properties of semiconductors it's very convenient to measure the energy in electronvolts and we will widely use this unit.

on the properties of the atoms comprising it. Depending on the material, it may vary from some tenths of fractions up to scores of electronvolts. (Thus, our rough calculation enables us to estimate the amount of the energy which is required.)

We cannot determine the distinct border and say that the substances whose value E_g is smaller than a certain value are called semiconductors, while those whose value is greater are called dielectrics. Though there were certain efforts to make such a classification, they might be compared to the scholastic questions how many hairs it takes to make a hairdo or how many grains it takes to make a pile of sand. On the other hand, when we say about some man that his hair is very thick and about the other man that he is quite bald, we very seldom meet with a difference of opinions. So keeping in mind that the figures we name are but relative and that the properties of semiconductors depend not only on the value of E_g, it is assumed that those crystals whose E_g is within some tenths of fractions up to 2–3 eV belong to semiconductors. Crystals whose $E_g \gtrsim 3$ eV in most cases behave as dielectrics.

When the value E_g is small, even a slight heating will cause the breaking up of a considerable number of electron bonds and the creation of a large number of free carriers. The crystal acquires an ability to conduct the electric current. It is clear that the substances, whose E_g is not large will take an intermediate place between conductors and insulators, since the ability to conduct the electric current, *conductivity*, is proportional to the concentration of the free carriers. In conductors (metals) the concentration of the free carriers is very great. It corresponds to the situation when all the electron bonds are broken and all the valence electrons are free. In "nonconductors" (dielectrics) there are practically no free carriers.

It is clear that resistivity of the material whose E_g is small is greatly affected by the temperature. In metals the concentration of electrons does not depend on temperature. Even at the absolute zero all electrons remain free and preserve their ability to conduct the current. In a typical dielectric, whose E_g is great, thermal motion cannot break the electron bonds and, consequently, the concentration of the electrons will not be affected by the temperature. In the semiconductor materials whose E_g is small the concentration of the free carriers, proportional to the number of the broken electron bonds, will grow very much with the increase of temperature. The resistivity ρ will accordingly be decreased.

Thus semiconductors are non-metal materials whose energy E_g is relatively small.

The value E_g for typical semiconductors varies from a few tenths of an electronvolt up to two or three electronvolts. Thus, for indium antimonide (InSb) $E_g = 0.17$ eV. Such small energy is possessed by quanta of infrared radiation whose wavelength $\lambda \approx 10$ μm. Therefore InSb is used to manufacture the radiation detectors which are essential in devices for night vision. For germanium (Ge), the material used to make the very first transistors and semiconductor diodes, $E_g = 0.72$ eV. For silicon (Si) which is the main material of the modern semiconductor electronics, $E_g = 1.1$ eV. Gallium arsenide (GaAs), the most promising material for semiconductor electronics of the near future, has $E_g = 1.4$ eV. In the ternary semiconductor compound GaAlAs, used to make semiconductor light-emitting diodes and lasers, the value E_g varies from 1.4 eV (GaAs) to 2.17 eV (AlAs), depending on the relative amount of Al and Ga they contain. In silicon carbide (SiC), the material of the most reliable and stable light-emitting diodes, able to operate at very high temperatures, $E_g \approx 3$ eV.

We have named here only the most important semiconductor materials. But hundreds of semiconductor compounds have been synthesized, studied and made use of.

"Free Electrons" and Holes

Now let us see what will happen if we break one of the electron bonds using the energy E_g, and releasing a free electron.

How can it be done? One of the ways is illuminating the crystal by using the light of a suitable wavelength. As we know, light is a stream of quanta of light (a stream of photons), each of them possessing a certain amount of energy $E_{ph} = h\nu = hc/\lambda$. In that equation h is Planck's constant equal to 6.62×10^{-34} Js, ν is the frequency, λ is the wavelength, c is the velocity of light, equal to 3×10^8 m/s. If the energy of a photon E_{ph} is greater than E_g, then a quantum of light is able to dislodge an electron from its orbit and thus create a free electron.* We can see in Fig. 3 what will take place

---

*It is convenient to measure the energy of a photon E_{ph} in electronvolts, and the wavelength of light λ in micrometers. Therefore the equation for calculating the energy of a photon may be written in the following way: $E_{ph} \cong 1.24/\lambda$. Derive this equation. Use this equation in order to find what kind of light is wanted: visible, infrared or ultraviolet – to create a free electron in Ge ($E_g = 0.72$ eV), GaP ($E_g = 2.3$ eV), C (diamond) – ($E_g = 5.6$ eV).

in this case. One of the links is broken (between atoms 19 and 20). The electron, being dislodged from its orbit, is within the area formed by atoms 1, 2, 6 and 7.

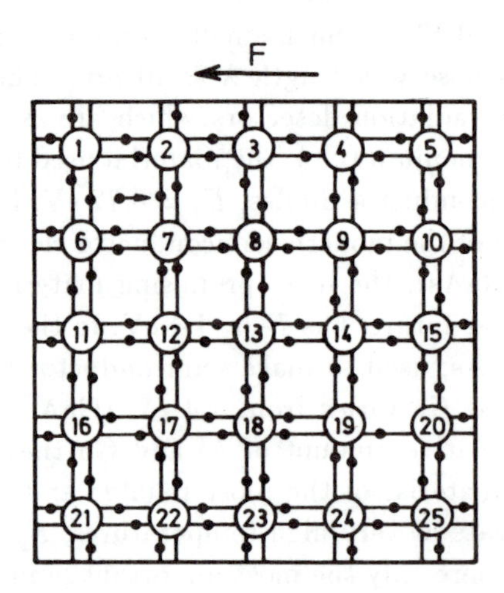

Fig. 3. The photon has dislodged the electron from the trajectory connecting atoms 19 and 20. There is an empty bond, a hole, and a free electron (between atoms 1, 2, 6 and 7). In the electric field F the electron moves to the right, and the hole to the left.

We can see that when speaking about an electron dislodged from its orbit we should preferably use the word "free", at least mentally, in inverted commas. The electron which is actually free is the one in the vacuum. While the so-called "free" electron in the crystal is in fact in a complex electric field. The electric field is formed by the ions of the lattice and by the valence electrons of the neighboring atoms. Under the action of the external electric field F a free electron in the vacuum moves with a constant acceleration $a = qF/m_0$ (m_0 is the mass of the free electron). The "free" electron in a crystal can move freely under the action of an external field only for a very short period of time, after which it is sure to collide with the atom of the lattice. Or else it might collide either with an atom of impurity or with a defect and so on. So, when using the terms "the free electron in a crystal", or "the free charge carrier", we should bear in mind that what they denote is not identical to the free electron in the vacuum. It denotes in fact just an ability to perform a directed motion under the action of the external electric field, thus conducting an electric current. That is why free electrons in a crystal are called *conduction electrons.*

Now it is high time to return to the other hero (should we say heroine?) of our "novel", the "heroine" that had been unattended by us for such a long time. We can see from Fig. 3 that one of the bonds between atoms 19 and 20 is broken, the electron having been dislodged from it. Now we will try to compare the motion of the dislodged electron and that of the vacant space, that is of the hole. If no external field is applied to the crystal, the electron will travel chaotically between the atoms of the lattice under the action of thermal oscillations. And what about the hole? Any of the electrons linking atoms 19 and 20 with the adjacent atoms may get to the trajectory of the dislodged electron and thus restore the broken bond between atoms 19 and 20. If that happened to be the electron which had been linking, say, atoms 14 and 19, then the broken bond (for the sake of brevity we will call it the hole) will take its position. Then it will be displaced to the position between atoms 9 and 14, or else between atoms 19 and 18; it may also move to the place between atoms 14 and 15. Just like the free electron, the broken bond, in other words the *hole*, will travel chaotically between the atoms of the lattice.

If the crystal is acted upon by an external electric field, then, apart from the chaotic motion, the free electron will also acquire a directed motion against the electric field. We must remember that the electron, being a negatively charged particle, moves from the "minus" to the "plus", i.e. against the vector of the electric field, which is always directed from the "plus" to the "minus". What about the hole? Let us look again at Fig. 3. Now any of the electrons, linking atoms 19 and 20 with the adjacent atoms, can shift to the trajectory between these two atoms (i.e. atoms 19 and 20). However, due to the action of the external field, this shift is most probable for the electrons linking atoms 18 and 19. They would be "pulled" by the external field to the place of the broken bond. The directed motion of the hole along the field will be added to its chaotic motion. Though it does not mean that the hole will be necessarily substituted by the electron linking atoms 18 and 19 and that the hole is sure to then appear between atoms 17 and 18. The external electric field does not stop the chaotic motion of the hole. It just adds to the chaotic motion some elements of the directed motion.

We must emphasize again that it is only electrons that are actually moving. The "hole" in fact is not a particle at all. It cannot be extracted from the crystal and studied in the vacuum. The hole merely denotes the absence of the electron in the bond linking the atoms of a crystal. It is however much easier to watch the displacement of a hole than the actual displacement of electrons from one orbit to another.

Let us make the following analogy (Fig. 4). Imagine that we stand on the balcony of a University building and watch the University square with lines of students, standing close to each other, shoulder to shoulder. The square is actually packed with students. All of them are wearing uniform caps. All of them but one sloven who has left his cap at home. Seeing that he is alone bareheaded and behaving a bit frivolously, which is not uncommon among students, he snatches his neighbor's cap and puts it on. Then his neighbor, in his turn, snatches somebody else's cap and so on. From above we can see the uncovered head – "the hole" – rapidly changing places, moving chaotically. Though we realize that the students do not move, they do not leave their places, that it is only the caps that are moving, it is much easier for us to watch the seeming motion of the uncovered head than to follow the actual motion of hundreds of caps. Let us give way to our imagination. Let us say that the Rector of the University suddenly comes out onto the balcony. Then... what a miracle! Under the Rector's strict gaze, "the hole", without stopping its chaotic motion starts shifting to the back rows. The directed motion is added to the chaotic motion. It is seen quite distinctly that "the hole" moves backwards, away from the Rector's strict gaze.

So, if we dislodge one of the electrons from its orbit (either by means of a quantum of light or in any other way), then the crystal will acquire not one but two opportunities to conduct the current:

1) The free electron moving under the action of the external electric field.
2) The electrons from the adjacent orbits moving towards the empty position in the link.

The second mechanism of conductivity can be defined as the motion of a "quasiparticle", i.e. of the hole, in the direction opposite to the direction of the electron motion.

Now let us discuss some properties of holes. The main question is whether we are to consider a hole to be a charged particle. On the face of it, it seems the answer must be negative since we have defined the hole as just

Fig. 4. The illustration of the notion of the "hole". It is much more convenient to watch one bare head than the actual displacement of many caps.

an empty place. But let us not make hasty conclusions. For when a crystal is acted upon by an electric field, then a break in the link is an additional source of conductivity. Electrons from the adjacent links jumping from one broken bond to another move against the field and conduct the current. If we are to define this mechanism of conductivity by means of the notion of a hole, we are to consider the hole, travelling in the field in the direction opposite to that of the electrons to be a *positively* charged particle whose charge is equal in magnitude to that of the electron.

The second question is whether we are to expect the electron and hole to have similar properties, e.g. the same speed of motion, when moving in the same electric field. But it is obvious that such an assertion would be groundless. Looking at Fig. 3 we can see that the conditions of the motion of a free electron within the crystal and the conditions of the motion of an electron from one bond to another are absolutely different. The motion from one orbit to another turns out to be much more difficult. In the same external field the hole, as a rule, moves more slowly than the electron.

And finally the last question. In the crystal which we have considered, electrons and holes are always created in pairs. The number of electrons is equal to the number of holes. Let the luminous flux which causes the appearance of the hole-electron pairs be continuous. Every photon which is breaking the bond, creates both an electron and a hole. Let us answer the following question. Can it be that even a weak light, illuminating the crystal for a long time, can break all the electron bonds and turn a dielectric into a metal? That would have happened if the newly created electron and hole live forever. But the lives of both of them, the electron and the hole, created by light are but very short. Their lifetime is different in different semiconductors but it is within the limits from 10^{-10} up to 10^{-2} s. How do they perish? In their chaotic motion in the crystal they may meet. Then the free electron takes its place on the free bond between the atoms and then ... both of them, the free electron and the hole disappear. This process is called recombination.

While recombination takes place, some energy E_g, which had been used to create the electron-hole pair is released. Sometimes recombination is accompanied by the birth of a photon and then there appears a quantum of light similar to that which perished while the electron and the hole were born. But it often happens that the electron and the hole recombine without giving birth to a photon. Then the energy E_g is wasted heating the lattice.

Thermal Generation of Electrons and Holes

A free electron and a hole or as they are sometimes called an electron-hole pair can appear in a crystal under the action of quanta of light, i.e. of photons, their energy $E_{ph} > E_g$. It is more or less clear that when the crystal is heated the atoms of the semiconductor are made to vibrate which results in breaking the electron bonds and in the appearance of electron-hole pairs.

What Will Happen if the Crystal is Heated? If we do not think of the quantitative relations, but consider only the physical picture we could be reasoning like this: the atoms of a semiconductor are known to have electron bonds with their neighbors. To break these bonds it is necessary to expand the energy E_g. When we raise the temperature of the crystal to such a level that the average energy of the thermal motion of the atoms will approach the value E_g, then we may expect the chaotic thermal motion of the particles to destroy the electron bonds. The crystal will then have free carriers, electrons and holes, and will acquire an ability to conduct the current.

This way of reasoning seems most verisimilar and convincing. Now let us verify it "numerically". The average energy of the thermal motion of particles is known to be equal to $3/2\ kT$, where k is Boltzman's constant and T is the absolute temperature. The value of $k = 1.38 \times 10^{-23}$ J/K or 8.6×10^{-5} eV/K. So, if that is correct, semiconductors will acquire an appreciable conductivity at the temperature T_0 which can be found from the equation: $3/2\,kT_0 \approx E_g$. Thus for InSb ($E_g = 0.17$ eV) the value $T_0 \approx 2000$ K, for Ge – 8400 K, for Si – 12700 K and for GaP – 27000 K!

Long before that temperatures might be reached, each of these substances is sure to be turned into vapor! Could we have made some mistake in our reasoning?

What Will Happen if We Place a Saucer of Water onto the Wardrobe? Let's check up once more our reasoning using a very common object. Let us pour some water into a saucer and put it onto the wardrobe out of reach of the cat. A few hours later when we look at the saucer we'll find that the water has evaporated. That will not surprise us at all, though, being physicists, we ought to be greatly surprised.

Indeed, room temperature is about 300 K. At this temperature the energy of the thermal motion of molecules of water is $3/2\,kT \approx 0.04$ eV.

Specific heat of evaporation of water at room temperature is 2.4×10^6 J/kg. Knowing how many molecules a kilogram of water contains and how many electronvolts one Joule is equal to, you may easily calculate that the energy of evaporation per molecule of water is $E \approx 0.46$ eV.* Thus, to break away one molecule from the surface of water in the saucer it is necessary to have the energy E more than tenfold greater than the average thermal energy. Nevertheless the molecules of water do leave the saucer!

The point is that besides the molecules whose energy of thermal motion is about kT (they comprise the greatest majority), water also contains such molecules whose energy at every given moment is much smaller than the average value and besides it also contains very "energetic" molecules whose energy is much greater. Every molecule acquires its energy at the account of accidental collisions with other molecules or with the walls of the vessel. A molecule experiences over 10^{12} (1000 billions) accidental collisions per second. No wonder it sometimes acquires energy much greater than the average energy. If that energy is more than 0.46 eV, then the molecule can fly away. The water evaporates.

If the saucer is covered, then between the surface of the water and the cover, there will be a certain concentration of molecules broken away from the surface of the water, the vapor. If the cover fits tightly, the molecules after they have wandered for some time between the cover and the surface of the water will return to the water. The quantity of the molecules leaving the surface of the water every second will be equal to the quantity of those which will return. There will be a state of dynamic equilibrium.

This is What Will Happen if a Crystal is Heated! Now we can see the mistake in our reasoning: we believed every particle to have an average thermal energy kT, while in reality there are particles whose energy is much greater than the average, including those particles whose energy is greater than E_g. A part of electron bonds between the atoms is broken: electrons and holes will appear in the semiconductor. The electrons and holes cannot leave the crystal. Therefore, having wandered for some time they will meet and recombine (i.e. they will disappear). A dynamic equilibrium will be established: as many electrons and holes will be created every second in the semiconductor, as many of them will perish in the process of recombination. There will be a certain equilibrium concentration of electrons and holes at a given temperature in the semiconductor.

---

*Try to calculate it.

In crystals whose properties we study, electrons and holes are always created in pairs and they always perish in pairs. Therefore the concentration of electrons n_i is equal to the concentration of holes p_i.* Let us determine the temperature dependence of the equilibrium concentration of the free carriers $n_i = p_i$.

To do it we must determine how many free carriers are formed in one unit of volume of the semiconductor per unit of time and how many of them perish during the process of recombination. In the case of equilibrium these two values must be equal to each other. Equating them to each other we will determine the equilibrium concentration of the carriers.

Thermal Generation. We have already learned the main feature of the electron-hole pair generation under the action of thermal motion. It takes much more energy than the average energy kT.

Investigating various physical processes we often come across such situations when to make a certain event take place it is necessary to expend a large "portion" of energy ΔE, much greater than kT. That happens when we study the evaporation of water, or the decay of the nucleus; when we investigate the emission of the electrons by the cathode of the vacuum tube or the distribution of the density of gases in the atmosphere. In one of the main parts of physics, statistical physics, it is proved that the probability of such an event is always proportional to $\exp(-\Delta E/kT)$:

$$w \sim e^{-\frac{\Delta E}{kT}} \tag{1}$$

where e is a number equal to $2,7183\ldots$ It is called the natural logarithmic base.

The dependence described by Eq. (1) is called exponential. Study it attentively; we will frequently come across it throughout the book, whenever we have to deal with events which need a much greater amount of energy than the average thermal energy kT.

The values of the energies $\Delta E = E_g$, necessary to break the electron bonds in various semiconductors are known to us. Consequently, knowing the temperature T, we can calculate by Eq. (1) the probability of forming an electron-hole pair in different materials. At room temperature ($kT = 0.026$ eV) for InSb ($E_g =$ 0.17 eV) this probability is proportional to $e^{-6.54} \approx 1.44 \cdot 10^{-3}$. For Si ($E_g =$ 1.1 eV) the probability is proportional to $e^{-42.3} \approx 4.23 \cdot 10^{-19}$; for GaAs ($E_g =$ 1.4 eV) $e^{-53.8} \approx 4.12 \cdot 10^{-24}$.

*The index "i" comes from the word "intrinsic".

Those examples illustrate the main feature of the exponential dependence: with the change of the index of the exponent, the value of the exponent changes a lot. The increase of ΔE by approximately 1.27 times diminishes the probability of the formation of an electron-hole pair approximately by 100 000 times! (cf. the probabilities for Si and GaAs). We will note that a similar change of the probability of forming electrons and holes will occur in one and the same semiconductor if the temperature is changed. So to decrease the probability of forming an electron-hole pair in Si by 100 000 times it is enough to cool the crystal of silicon, lowering the temperature from room temperature to -78°C, i.e. to the temperature of the dry ice, used to preserve ice-cream.

So, the number of electron-hole pairs K_1, being formed every second in a unit of volume of a semiconductor, is

$$K_1 = \alpha e^{-\frac{E_g}{kT}} \tag{2}$$

where α is a coefficient of proportionality, different for different semiconductors.

On the other hand, due to recombination, a certain number of charge carriers K_2 will disappear from the same unit of volume every second. So what does the number of recombining carriers depend on?

Electron-Hole Recombination. In order to recombine, the electron and hole must meet. What does the frequency of their meetings depend on? Let us mentally make the following experiment. We will watch some atom of the crystal lattice and mark the appearance of a hole in its vicinity. It is clear that the greater the number of holes per unit of volume of the semiconductor, the more frequently they will appear there. The probability of the appearance of a hole is proportional to the concentration of holes p_i. For the same reason the probability of the appearance of a free electron in the vicinity of that atom is proportional to the concentration of electrons n_i. We are interested in the probability of a simultaneous appearance of both the electron and the hole in the vicinity of the atom. This probability (i.e. the probability of their meeting and recombination) is proportional to the product of the concentrations of electrons and holes $n_i \cdot p_i$. Thus, the number of carriers recombining every second per unit of volume is the following:

$$K_2 = \beta n_i \cdot p_i \ . \tag{3}$$

The coefficient β as well as α in Eq. (2), is different for different semiconductors.

Since $n_i = p_i$, then

$$K_2 = \beta n_i^2 = \beta p_i^2 . \tag{4}$$

Equating the number of the newly born pairs K_1 to the number of the perishing pairs K_2, we obtain:

$$\alpha e^{-\frac{E_g}{kT}} = \beta n_i^2 = \beta p_i^2 . \tag{5}$$

Hence we obtain

$$n_i = p_i = \sqrt{\frac{\alpha}{\beta}} e^{-\frac{E_g}{2kT}} = (A \cdot B)^{1/2} e^{-\frac{E_g}{2kT}} . \tag{6}$$

It is worthwhile discussing Eq. (6) in detail.

Intrinsic Semiconductors

First of all it should be mentioned that a semiconductor whose concentration of carriers is determined by Eq. (6) is called *intrinsic.* The intrinsic concentration of electrons and holes, $n_i = p_i$ is determined by temperature T and energy E_g, the most important characteristic of a semiconductor.

Values A and B in Eq. (6) are measured in cm^{-3}. They are well known for all the semiconductors we are to deal with. At room temperature these values are within the limits approximately from 10^{17} to 10^{19} cm^{-3}. Knowing the values of A and B, it is easy to calculate by Eq. (6) the values of the concentration of intrinsic carriers in any semiconductor at any temperature. Table 1 gives the values E_g and n_i for some semiconductors at room temperature (300 K).

Table 1

Semiconductor	InSb	Ge	Si	InP	GaAs	GaP	SiC
E_g, eV	0.17	0.72	1.1	1.3	1.4	2.3	2.4–3.2
Intrinsic concentration n_i, cm^{-3}	$1.3 \cdot 10^{16}$	$2.4 \cdot 10^{13}$	$1.1 \cdot 10^{10}$	$5.7 \cdot 10^{7}$	$1.4 \cdot 10^{7}$	0.8	$0.12–2 \cdot 10^{-8}$

What conclusions can be drawn from this Table? First let us put a simple question: is it easy to study electric properties of intrinsic semiconductors? One cubic centimeter of various crystals contains $\sim 10^{22}$ atoms. A substance is usually called pure if it contains one foreign atom per 1000 intrinsic atoms (i.e. the impurity concentration $\sim 0.1\%$). From the chemical point of view the substance will be absolutely pure if it contains one foreign atom per 100 000 intrinsic atoms. Nevertheless it means that every cubic centimeter of the substance will contain $\sim 10^{17}$ foreign atoms. And now let us imagine that there is an impurity in the semiconductor which is able to release quite easily a free electron or form a hole. (A bit further we will see that very many impurities possess this property). Then germanium, absolutely pure from the chemical point of view, (the concentration of impurity will be only 0.001%) will at room temperature contain such a number of impurity electrons which will exceed the number of intrinsic electrons by a factor of 4000. In silicon the number of impurity electrons will be 10 000 000 times greater than that of the intrinsic ones: in gallium phosphide it will be 10^{17} times greater! But the initial materials used to produce a semiconductor crystal also contain impurities. Besides, there are impurities in the walls of the furnaces and installations where the synthesis and purification of semiconductors take place and in the atmosphere as well.

It is absolutely impossible to get rid of impurities. Though due to the joint efforts of physicists and technologists, of metallurgists, chemists and a lot of other specialists, there have been various methods devised to produce semiconductor crystals of wonderful purity.

SUMMARY

Semiconductors are such materials whose conductivity at the temperature of absolute zero is equal to zero, all the valent electrons being bound in the interatomic orbits and thus unable to conduct an electric current. But the energy of the electron bonds is not very great, and even when the temperature is not high, the electron bonds break down on account of the thermal motion. There appear free electrons (conduction electrons) and holes. Their concentration increases exponentially with the increase of temperature. And accordingly, with the rise of temperature, the conductivity of semiconductors increases exponentially too.

Part I. THE HISTORY OF SEMICONDUCTORS

> ... Who believes, that in all ages
> Every human heart is human,
> That in even savage bosoms
> There are longings, yearnings, strivings
> For the good they comprehend not,
> That the feeble hands and helpless,
> Groping blindly in the darkness,
> Touch God's right hand in the darkness,
> And are lifted up and strengthened
> Listen to the simple story ...
>
> H. Longfellow
> 'The Song of Hiawatha'

"Semiconductors is a comparatively young science, it isn't thirty as yet. It has the features of the new times." Those words were written in 1956 by Abram F. Ioffe, a Russian academician.

"The English physicist Cavendish has proved experimentally that water conducts electricity 400 million times worse than metals; nevertheless it is not a very bad conductor of electricity. Bodies which take the intermediate position between conductors and nonconductors are usually called semiconductors." These words were written in 1826 by Ivan Dvigubsky, the author of the textbook *"Fundamentals of Experimental Physics."*

The last quotation is remarkable in many respects. Firstly, it is clear from it that as far back as 150 years ago there were materials which were "usually called" semiconductors. Secondly, that they were defined by that very property which we have begun this book with.

The most thrilling part of the quotation is the following: "... Cavendish has proved experimentally that water conducts electricity 400 million times worse than metals..." On the face of it the sentence looks most natural. But what experiments are meant here? How could Cavendish, the famous physicist and chemist of the 18th century, compare the conductance of metals, dielectrics, and water?

Let us give way to our imagination and visit the ancient castle of one of the richest people of England. Henry Cavendish, wearing an ancient robe and silk stockings works at his ancient writing table (Fig. 5). There is an ancient galvanometer, an ancient voltmeter and an ancient accumulator on the table. By means of those things Lord Cavendish measures the conductance of water.

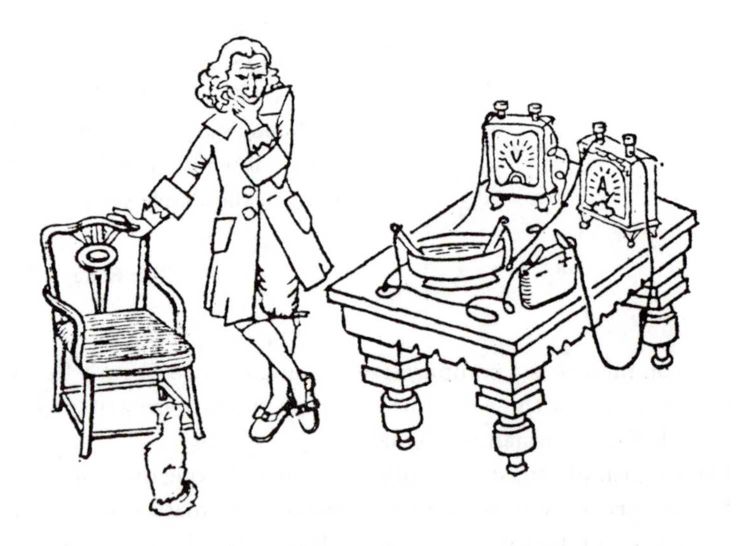

Fig. 5. Lord Cavendish is thinking over his experiment.

Cavendish made his experiments on measuring the electric resistance in 1776. If we put this date under the picture we have drawn, and show it to a specialist in the history of physics, his reaction would be unpredictable. I wish he were a man of nerve and had a sense of humor. Otherwise...

The first accumulator was invented by the Italian physicist A. Volta in 1800, i.e. 24 years later. (So, the way it is done in the cartoons, the accumulator disappears – there isn't any accumulator on the table).

The first galvanometer was made by the German physicist Wilhelm Weber in 1846, i.e. 36 years after the death of Cavendish. (So, both voltmeter and galvanometer disappear from the table).

Now, in what way could Lord Cavendish prove anything experimentally at all? How could he know in 1776 that there are metals, dielectrics, and semiconductors? How was it possible then to study the electric properties of materials?

Chapter 1. *DE PROFUNDIS* (FROM THE DEPTHS)

Do not listen to those thinkers, whose arguments are not proved by experiments.

Leonardo da Vinci

The physics of semiconductors is a fruit of a delicate, inconspicuous branch of a powerful ancient tree of the science of electric phenomena.

THE GENIUS OF ANCIENT GREEKS

Amber, rubbed with wool, acquired the property of attracting light objects – specks of dust and straws. The honor of this discovery has been ascribed to the Greek philosopher Thales from the town of Mileta, who lived in 640–550 B.C. Thales believed amber to have a soul, which was the source of attraction, because only something animate could cause motion.

The term "electricity" is derived from the Greek word "electron" which means amber. The Greek did not know yet that other substances, apart from amber, could also be electrified when they were rubbed. Though Theophrastos, who lived 300 years later than Thales, mentioned another substance: lincurion. He wrote about it in his composition, "On Stones", describing it as another electrifying stone. But nothing is known to us about that stone.

There is a fish in the Mediterranean, the skate, which was called "narcae" by the ancient Greeks which means "paralysing". It is known nowadays that the voltage generated by this fish reaches 200 V. The Roman physician Scribony, who lived at the beginning of our era used narcae to treat headaches, gout and some other diseases without knowing anything about the electric nature of this method. Electricity, generated by the skate was the first natural electric phenomenon used by man, though Scribony thought it was a certain poison which had the healing effect.

As for amber, its ability to attract specks of dust caused irritation rather than interest: the expensive beads of ladies were always covered with specks of dust...

Life went on, years and centuries passing by. New civilizations would come into existence, would flourish and disappear... Very slowly, overcoming a lot of hardships, mankind learned to perceive the laws of nature and to make use of them.

In the 2nd century B.C., Archimedes formulated the laws of floating of bodies and created the fundamentals of mechanics. In the 3rd century A.D., in China appeared the first compass. In the 11th century Muhammed ibn Ahmed al-Biruni, a famous oriental scholar of the Middle Ages, measured the densities of many substances and determined the radius of the Earth. In the 15th century, Iohann Gutenberg invented book-printing and in the 16th century Nicholas Copernicus and Jordano Bruno proved the rotation of the Earth...

For 2000 years not a single discovery had been made in the field of electrical phenomena.

WILLIAM GILBERT (1544–1603)

In the last years of the 16th century William Gilbert, physician to the British Queen Elizabeth Tudor, Doctor of Medicine and Master of Arts, famous for his knowledge of chemistry, physics and astronomy, devoted himself to studying magnetic phenomena. The result of his experiments and reasoning was his composition, "On Magnet, Magnetic Bodies and on the Big Magnet – the Earth". It was published in 1600 and comprised six volumes. The second volume of this composition described the experiments in the field of electricity. Using a very simple device, the prototype of the future electroscope (Fig. 6), Gilbert showed that not only amber, but also diamond, sapphire, amethyst, rock crystal, sulphur and some other

William Gilbert was a most outstanding English scholar of the Renaissance period. He was born in 1544 in the town of Colchester, where his father was the town's judge. William was a student of Oxford and Cambridge and then became a practising physician. At the age of twenty he obtained his Bachelor of Medicine and Master of Arts degrees, at twenty-five he was Doctor of Medicine.

William Gilbert
(1544–1603)

Gilbert was not only a skilled doctor and a successful physician to her Majesty Queen Elizabeth Tudor. He was also a brave and gifted scientist, one of those who supported ardently the theory of Copernicus. He was one of the founders of a new natural science, in which experiment was the main criterion of truth.

Gilbert was the first naturalist whose approach to the study of electric and magnetic phenomena was truly scientific. He was the first to understand that it was the Earth itself which was the source of attraction for the compass needle. He was the first to introduce the notions of the magnetic axis, magnetic meridians and parallels. He found that the magnetic properties of iron disappear when the iron is heated. He established that almost 300 years before the work by P. Curie.

The work by Gilbert marked the birth of a new science, the science of electricity.

Gilbert died in 1603, being infected with plague. He bequeathed his large collection of books as well as the devices created by him to the College of Physicians, but the greater part of those materials were burned down during the fire of London in 1666.

The most remarkable work by Gilbert is "On Magnet, Magnetic Bodies and on the Big Magnet – the Earth". The work has been preserved and has become one of the most interesting documents of 17th century science.

substances which he called "electrical", when rubbed, acquired the property of attracting the needle of the device. He also discovered that pearls, marble, bone and metals do not acquire that property. It does Gilbert credit that he insistently popularized the advantages of the experimental researches over sophistical exercises and verbose "beating the air". "If someone does not agree to some opinion or paradoxes..." Gilbert wrote, "... let him pay attention to the numerous experiments and discoveries that have been made, study them most attentively, and much good may it do them!"

Gilbert's experiments were continually repeated in many countries.

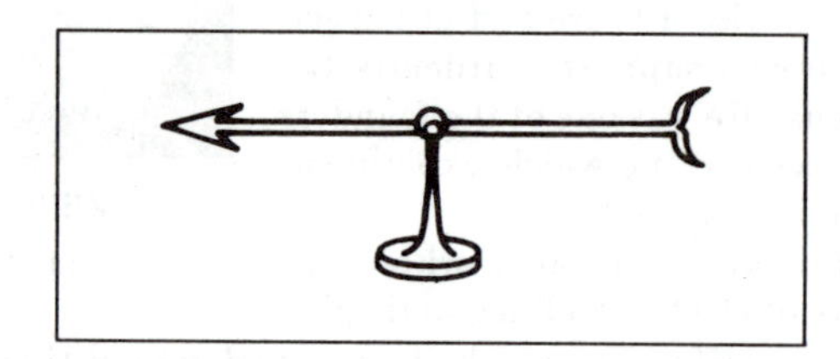

Fig. 6. Versor – the prototype of the future electroscope. It was apparently invented by Girolame Fracastere (Italy) in the middle of the 16th century. Even a small force makes a light arrow rotate around an axis.

GERMANY, ENGLAND, FRANCE AND AMERICA: OTTO GUERICKE, STEPHEN GRAY, CHARLES DUFAY, AND B. FRANKLIN

In 1672, an important discovery was made by Otto Guericke, burgomaster of the town of Magdeburg. He made a ball of sulphur whose diameter was about 15 cm. The ball was fixed on a metallic spindle and was rotated with a handle and when rubbed with the palm of the hand gave much stronger electricity than in the experiments made by Gilbert. It was the first electric machine.

At the beginning of the 18th century, Stephen Gray, a member of the London Royal Society, repeating the experiments made by Gilbert, noticed that bits of fluff were attracted not only by a rubbed glass tube but also by a ball made of ivory and fixed to the tube by means of a hempen rope. This phenomenon interested Gray and he began to increase the length of the rope. At the length of 26 feet (approximately 8.5 m) the ball was still attracting bits of fluff. Gray had to climb onto the roof – the attraction was as strong as if the ball itself were electrified. But when Gray substituted a silk thread for the hempen rope the attraction disappeared. Gray understood that unlike hemp, silk did not "conduct" electricity. Then he substituted a metallic wire instead of the silk thread, the ball was again electrified by the glass tube.

Gray's report at the Royal Society aroused much interest and stimulated some new research. The French scholar Charles Francois Dufay repeated that experiment and came to the conclusion that not only those substances which were mentioned by Gilbert but practically all the substances except metals and humid bodies are electrified when they are rubbed (Fig. 7).

Fig. 7. Charles Francois Dufay demonstrates a phenomenon of electrization (18th century). His heels in silk stockings are electrified by a rotating glass ball. A spark jumps between his nose and a lady's finger.

In 1739, J. T. Desaguliers, professor of physics and theology of Oxford University, repeated these experiments and introduced the terms "conductor" and "insulator" which showed the properties of the bodies to conduct or not to conduct electricity.

Electric phenomena interested not only scholars. Thanks to the Guericke electric machine and to some other similar devices, electric phenomena were then demonstrated everywhere: in the squares, in homes and even at the Court. Wandering jugglers began to profit from demonstrating various experiments with electricity: the thrilled audience watched the spirit in a bowel catching fire from the spark, coming from the fingers of an electrified madam. Electricity became very popular. It was considered good for the health to drink the electrified water, so they plunged the end of the wire from some electric machine into a jug of water before they drank it.

Sparks, luminescence and crackling accompanying the work of the electric machines made curious scholars look up doubtfully at the sky. Could it be that lightning, that "fiery arrow of Zeus" was a gigantic spark of some monstrous natural electric machine? The French scientist Dufay made a supposition which was rather bold for his time (1735): "In the long run we may be able to obtain electricity on a large-scale and then we will amplify the electric fire whose nature might be similar to that of thunder and lightning."

And now let us travel across the ocean to North America. Benjamin Franklin, a bright politician and diplomat, one of the leaders of the struggle for the independence of the USA, one of the authors of their first Constitution and of the petition on abolishing slavery, one of the most popular writers of his time, at the age of forty became keen on research in the field of electric phenomena. He made many important discoveries in this field and established the electrical nature of lightning.

He also, just like Dufay, paid attention to the likeness between the spark and the lightning, between the crackling of the electric machine and the roars of thunder. Franklin's main experiment which enabled him to establish the nature of lightning was the work of a genius, it was quite simple. When low clouds covered the sky and the storm was expected to break, Franklin launched a kite made of paper.

The kite was equipped with a metallic point, and a key made of metal was fastened to the end of the rope near the ground. The rope was wet from the rain and when metallic objects were brought up to the key the latter emitted electric sparks followed by crackling. The natural electricity,

Benjamin Franklin
(1706–1790)

Benjamin Franklin was born in January 1706, in Boston. He was the eighth child in the family. His father, Joseph Franklin, an emigrant from England, was a handicraftsman manufacturing soap and candles. There were 17 children in the family, so Benjamin had to begin working when he was ten.

Franklin's brightness and various abilities were displayed very early. He did everything he could to educate himself and spent every cent he could spare on books. After having worked in a number of workshops in Boston and New York, Franklin moved to Philadelphia where he organized a print-shop. When he became rich he devoted much time to social work and not leaving it towards the end of his life.

Franklin's natural gifts were manifested best in his research work. In 1746, he happened to be present at the show "Wonders of the Physics Study-Room"organized by Dr. Spens. There Franklin saw for the first time the electric machine and was greatly interested in it and in the experiments in electricity. Franklin made several remarkable discoveries in the science of electricity. He explained the effect of the Leyden jar; he designed a flat capacitor whose insulation was made of glass; he discovered the electric nature of lightning and was the first to invent a lightning rod.

Franklin was elected Member of the London Royal Society and was awarded its highest prize – the Copley medal. In 1789, Franklin was elected honorary member of the Russian Academy of Sciences.

America of the 18th century did not know any man as widely gifted and industrious as Franklin. Being a writer and a publisher, a scientist, a public worker and a diplomat, he was called "the first among the civilized people of America".

Franklin died on April 17, 1790 at the age of 84. America was in deep mourning for thirty days. By the decision of the World Council, the name of Benjamin Franklin was listed among the names of the best representatives of mankind. "His genius manifested the beauty of human nature and his aspiration for the benefit of mankind did a lot to his motherland, science and liberty".*

*From the resolution of the U. S. Congress on a thirty-day long mourning on the account of the death of Franklin.

which was accumulated in the clouds caused exactly the same phenomena that was already known to researchers working with the electric machines.

The practical American suggested a lightning conductor – a grounded metallic rod, isolated from houses and raised high above the ground. Nowadays, when such devices are quite commom, it is hard to imagine what hot discussions they caused in those times.

Franklin's contribution to the research of the first electric capacitor, the "Leyden jar" was also very great.

THE HISTORY OF THE LEYDEN JAR

In the years 1745–1746, Urgen von Kleist, dean of one of the cathedrals of Pomerania, and the professor of philosophy from Leyden Musschenbrock almost simultaneously made a wonderful discovery. This is what Musschenbrock wrote to one of his correspondents: "I want to inform you of my new but awful experiment which I would not advise you to repeat. I tried to study the electric force, so I had a metallic tube, suspended on two blue silk cords while the tube was electrified by a glass ball which was rubbed between the palms of the hands and was rotating with a great speed. At the other end there was a copper wire the end of which was plunged into a round bowl with my right hand while with my left hand I tried to get some sparks from the electric tube. Suddenly my right hand was struck with such a force that all my body shuddered as if hit by lightning... I thought it was the end."

But no one followed Mussenchenbrock's advice not to repeat that experiment. On the contrary, it caused extraordinary interest and made many people of different social standing continue such experiments.

His majesty, the King of France, Louis XV sent the discharge of the Leyden jar through a line of soldiers and enjoyed himself watching the grimaces of the poor wretches. And one of the leaders of the French revolution physician Jan Pole Marat wrote a composition "How Much and Under What Conditions We May Rely on Electricity in Treating Diseases". For that work he was awarded in 1783 the Gold Medal of the Rouen Academy. But physicists benefited most of all from the discovery of the Leyden jar. Now they had at their disposal a powerful and reliable source of electricity.

In 1761, Franklin's friend Ebenezer Kennerley managed to bring the conductor, through which a Leyden jar discharge was passed, to red heat.

A few years later, Joseph Priestly noted that the melting of different metals caused by a spark could be used as a criterion of their electroconductivity.

That was proved experimentally by Jambatist Beccaria in Italy and by John Canton in England. And at last one of the greatest scientists of the 18th century, Henry Cavendish, made a systematic measuring of the relative resistivity of the various substances. And that was in 1776, i.e. 50 years before the discovery of Ohm's law!

Now coming back to Fig. 5 we can remove from the ancient table the accumulator, voltmeter and galvanometer and put there the devices which were actually used by Cavendish in his wonderful experiments: the Leyden jars and the electric machines to charge them (Fig. 8).

"He used to take two wires of different metals, the wires had the same length and the same cross section, and discharged through them the Leyden jars. He would repeat that experiment changing the charge of the battery until one of the wires melted. Assuming that the wire which did not melt had a higher electric conductivity, he obtained a row of metals arranged according to their conductivity."

It is simple, isn't it? But one had to be a genius to think of it.

METALS, DIELECTRICS, "SEMICONDUCTORS"

Thus we have seen how in the course of 180 years, researchers who studied electricity have arrived at the idea of conductors and insulators. And also of "semiconductors" – substances about which nothing definite was yet known. They did conduct electricity so they could not be defined as insulators, but they did not conduct it well enough to be defined as conductors. It was hard to expect this group to be of much interest for the researchers.

THE 19TH CENTURY: ITALY, DENMARK, FRANCE, GERMANY: VOLTA, ØRSTED, AMPERE, SEEBECK

The beginning of the 19th century was marked by a wonderful discovery made by the Italian physicist Alexandre Volta. In 1800, Volta reported that a column made of alternating discs of silver and zinc, separated by wet cardboard circles, could be used as a source of direct current for many hours. This discovery allowed scientists to watch many phenomena which

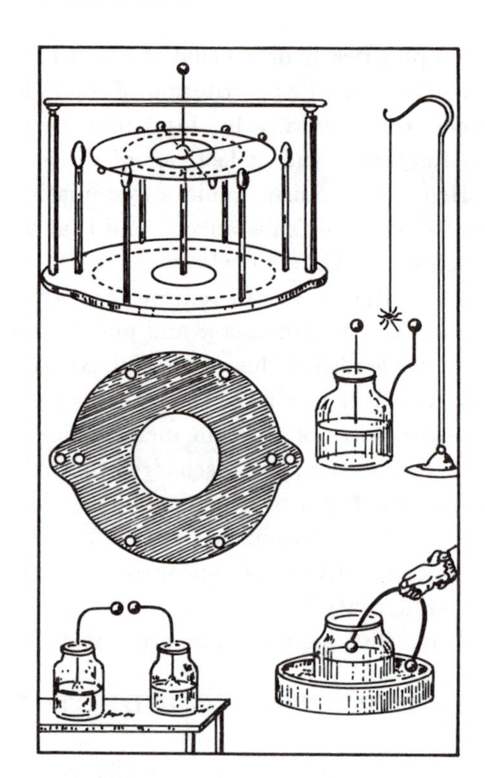

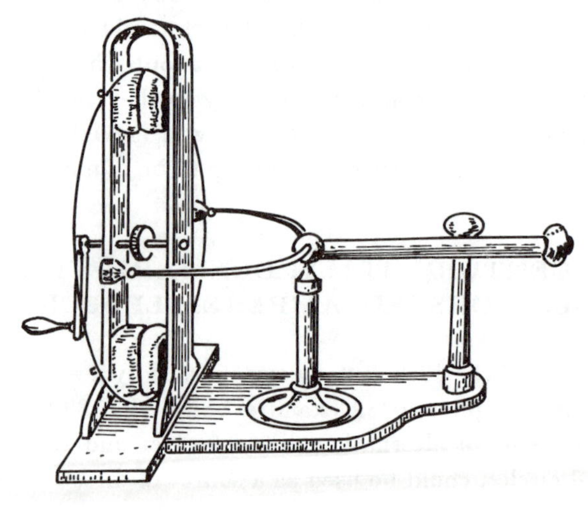

Fig. 8. Leyden jars and electric machines.

Henry Cavendish. This famous English chemist and physicist was born on the 10th of October, 1731. He was a wonderful man of an unusual fate. He was the eldest son of Lord Charles Cavendish and belonged to a very noble, though not very rich family.

Henry Cavendish
(1731–1810)

At the age of forty, Henry Cavendish inherited an immense fortune and became one of the richest men of England.

He spent the greater part of his inheritance on organizing scientific experiments and helping young scientists. He avoided all kinds of society and saw very few of his relatives. Deep in meditation, trying to save time for his research work, this most unusual man used gestures rather than words when addressing people at home.

Cavendish was a genius. He anticipated many outstanding discoveries in physics and chemistry. He became famous for his investigation of the properties of gases. He was the first to obtain hydrogen by adding hydrochloric acid to zinc (and to iron), and he proved that water is a compound. He determined the amount of oxygen in the air. Cavendish discovered the law of the interaction of electric charges 12 years before Coulomb, though he did not publish his results anywhere. He measured the resistivity of many substances long before the discovery of Ohm's law.

Cavendish computed and manufactured meteorological devices, devices for making the chemical analysis of gases, he created apparatuses for navigational astronomy and so on. His motto was: "Everything is defined by measure, quantity and weight." The precision of his experiments is just surprising. And that was in those times when there were practically no measuring instruments at all. The law of interaction of electric charges was verified by him to a precision of 0.025%, the amount of oxygen in air was found to be equal to 20.83% (the modern value is 20.93%), the density of the Earth – 5.45 g/cm^3 (by modern data it is 5.52 g/cm^3)!

Cavendish died in 1810 in London.

The numerous works by Cavendish had not been published by him, especially in the field of electricity. They had been kept as manuscripts for more than 100 years until 1879 when J. C. Maxwell studied and published them. It was only then that they became widely known and highly appreciated.

could not be observed during a momentary passing of a spark, from the Leyden jar or from an electric machine.

In 1819, the Danish physicist Hans Christian Ørsted, while using the Volta source noticed the deflection of the magnetic needle if it was near the conductor along which a current was flowing. This discovery as well as the brilliant experiments made by the French scholar Andre M. Ampere, which demonstrated the close connection between electric and magnetic phenomena, gave rise to thousands of experiments. One of them which was not considered sensational at that time appears very interesting now.

In 1821, the German physicist Tomas Seebeck, impressed by the experiments made by Ørsted and Ampere, also decided to investigate "the magnetic sphere of electric current". A sketch of his experiment is given in Fig. 9.

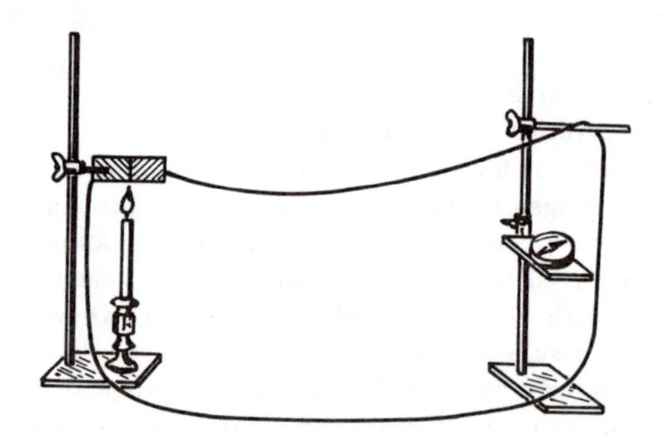

Fig. 9. Seebeck's Experiment.

Seebeck brazed two different metals to each other, joined them to a copper conductor and placed a magnetic needle inside the loop formed by the conductor. Heating the brazing by a candle, Seebeck noticed that the magnetic needle located close to the conductor, deflected. Heating the joint of two different metals generated an electric current! Seebeck investigated this effect most thoroughly and efficiently, experimenting with hundreds of different materials. He found out that in the case of one of the elements of the joint being either tellurium or lead sulphur, or else some other materials (which 100 years later would be called *semiconductors*), the deflection of the needle increased very much. In the course of many years the current sources based on Seebeck's effect were widely used in experimental physics. In order to increase the strength of the current the researchers connected many tens or even hundreds of joints (thermocouples), making constructions called thermopiles (Fig. 10).

It would be natural to expect that for the elements of these thermopiles such substances would be used which would give the maximum strength of current. This would have resulted in a better study of the properties of such substances, but as the Greek saying goes "Even Gods can't change the past".

The thermoelectric sources were appreciated first of all from the point of their stability and reliability. It was much more simple to double the number of the metallic layers than to try to find why in one case using the lead sulphide gave wonderful results and in another case the results were useless. In the second part of the book we shall get acquainted with the striking sensibility of the semiconductors to the impurities and contaminants and then we will realize what kind of "stability" or "purity" of semiconductors could be expected as far back as 150 years ago.

The first signal of the existence of such a class of substances, possessing quite peculiar properties, remained unnoticed.

The second signal was spotted by the great English physicist Michael Faraday.

Faraday's teacher, the famous chemist Humphry Davy, established as far back as in 1821 that conductance in metals decreases with the increase of temperature. In 1833 when Faraday continued Davy's experiments on investigating conduction-temperature dependences of different substances he was faced with an unusual situation. He found that the conductivity of silver chloride, which at that time was considered to be a metal, did not fall, but increased with the increase of temperature. Faraday described the results of this experiment in his famous "Experimental Research in Electricity": he wrote that he did not know any other substance which like silver sulphide, when heated, could be compared with metals with regard to their conductivity, but whose conductivity unlike that of metals decreases upon cooling, while the conductivity of metals increases with cooling. He thought that if they searched for such substances they might find them.

In the course of the next five years, Faraday did discover a few more "such substances", lead fluorine PbF_2, mercuric sulphide HgS, and some other substances also possessing that unusual conductivity-temperature dependence. In 1838, when Faraday had completed measuring the conduc-

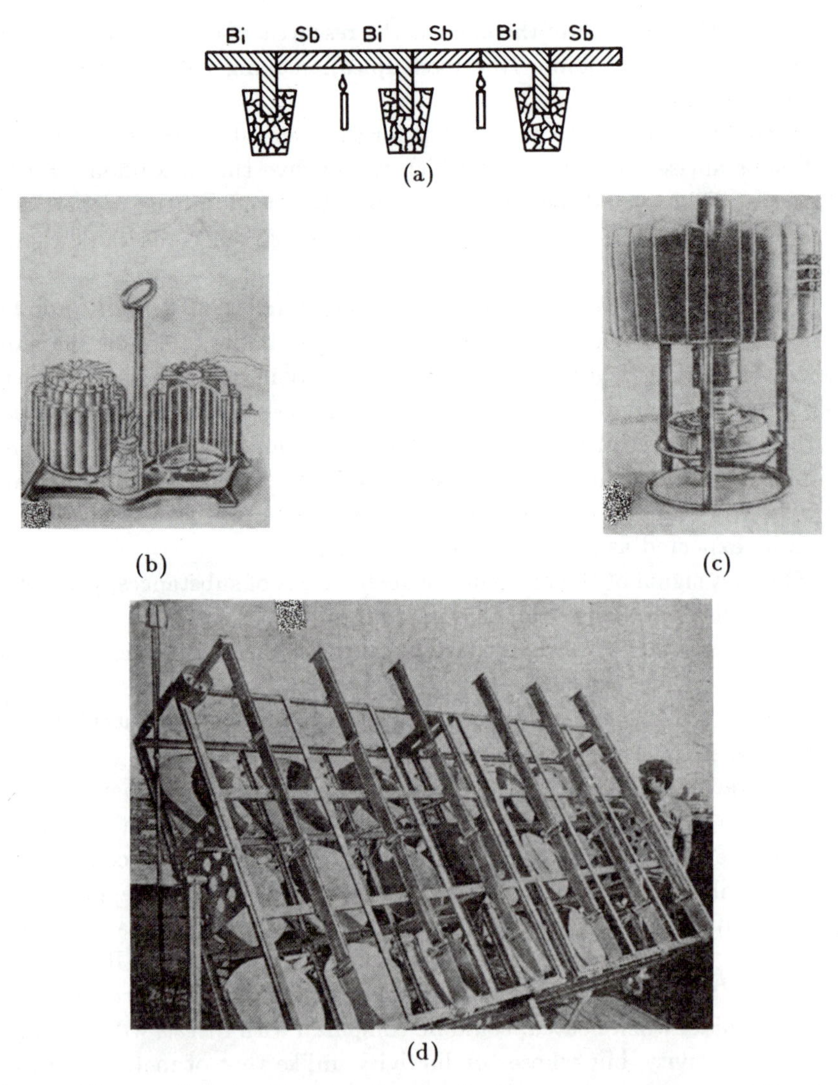

Fig. 10. Thermopiles. (a) The first thermopile (invented by Ørsted and Fourier). The hot contacts of bismuth (Bi) and antimony (Sb) are heated by candles. The cold seams are plunged into vessels with ice. (b) The Neah cell (the seventies of the 19th century). Hot seams are heated by a Bunsen burner. Efficiency – 0.1%. (c) Thermopile (the fifties of the 20th century). Hot seams are heated by a kerosine lamp. Efficiency $\sim$ 3%. (d) Modern solar battery. Concave reflectors concentrate solar radiation on the cells, whose base is the contact of gallium arsenide with GaAlAs – a semiconductor compound. Efficiency $\gtrsim$ 20%.

Michael Faraday
(1791–1867)

Michael Faraday. Michael Faraday was born on September 22, 1791 into the family of a poor London smith. He had to begin working at the age of twelve. First he was a bookbinder's apprentice, then an errand boy at a bookshop. He acquired deep knowledge of chemistry and physics, studying those subjects independently and most insistently. Faraday was 19 when he was offered tickets for a course of lectures to be delivered by Sir Humphry Davy. That event predetermined his fate. The young man attended all the lectures, listened to them attentively and made notes. Then he bound the records he had made and sent them to Sir Humphry Davy with a letter which he enclosed with the records. That resulted in his being accepted to the Royal Institute as an assistant.

Faraday was 22 when together with Sir Davy, he made his first experiments on electricity. They investigated the powerful pulses made by the skate. To the last day of his life, apart from working at a lot of diverse problems, Faraday had kept puzzling over the mystery of electricity.

Michael Faraday is one of the most gifted and famous physicists of all times. He is the author of numerous wonderful pieces of work in the fields of physics and chemistry. Among his discoveries are the following: the laws of electrolysis, the law establishing a connection between the magnetic and electric phenomena, the discovery of dia- and paramagnetism and the discovery of the rotation of the light polarization plane in the magnetic field.

Michael Faraday was the first to discover and prove experimentally the effect of an abrupt increase of conductance as a function of temperature for a number of substances (sulphide silver, lead fluorine, etc.). Faraday believed that there must be some deep physical grounds for that effect which might perhaps indicate the existence of a new class of substances.

Nowadays these substances are called *semiconductors*.

Faraday died on the 25th of August, 1867. Scientists throughout the world revere the man. They revere him not only as a man of genius, a wonderful experimental physicist and educator, but also as a man of scrupulous honesty, modesty, and conscientiousness.

tivity-temperature dependence, he was famous all over the world for his discovery of the laws of electromagnetic induction and electrolysis. He was at that time a member of the London Royal Society, Director of the Laboratories of the British Royal Institute, honorary professor of chemistry. All his results were regularly reported and published by him. But though his contemporaries knew the results of his experiments and his forecasting the probable existence of a class of peculiar substances with specific electric properties, no further work was done in this field. Faraday's appeals were not answered. The second signal was not attended to either.

Was it just accidental? Or was it a bitter paradox of history?

Nothing of the kind. It was just normal. As it was said by James C. Maxwell, a distinguished physicist of the previous century, it is the great need of practical knowledge that makes any branch of natural sciences develop quickly and fruitfully.

With regard to semiconductors this need of knowledge was felt a few decades later. It happened 40 years later as a result of the development of... the distant telegraph communication.

SEMICONDUCTORS ARE WANTED EVERYWHERE IN THE WORLD

The different versions of the telegraph apparatus which could transfer information along large distances were rather numerous after Ørsted's discovery of the magnetic action of the electric current. The first apparatus was constructed by the Russian inventor P. L. Shilling. Electromagnetic telegraphs were built in the thirties of the last century in Germany, England and in America.

In 1844, the New York artist Morse, who was fond of electrotechnics, invented an apparatus with a moving paper tape on which a pencil attracted by an electromagnet was leaving dots and dashes. This invention marked the turning point in the history of the mail; throughout the world telegraph began to substitute mail-coaches. By the seventies of the 19th century the telegraph was everywhere. There was a network of telegraph companies, there appeared a new speciality of an engineer-telegrapher, and new societies of telegraphers were formed.

In 1873, an electrical engineer from London, W. Smith investigated the subaqueous cable. He decided to use selenium to insulate the cable. Selenium when melted and then cooled very quickly, becomes a vitreous mass whose resistance is very great. This mass was used as an insulator for that cable. Smith's assistant, May, who watched the experiment very attentively, noticed that in broad daylight the resistance of selenium becomes much smaller than in the darkness. He reported this phenomenon and it gave rise to a series of experiments. This discovery was proved by not fewer than a dozen physicists. It was discovered that selenium was sensitive even to the dim light of the moon.

Selenium ability to transform light into electric signals answered a lot of practical needs. Photoconductors made of selenium were widely used in various optical devices.

It became necessary to explain the reason of the photosensitivity of selenium. They began to search for the new photosensitive materials. At the beginning of the 20th century physicists began to study specially the materials which were neither metals nor dielectrics.

Then they remembered the extraordinarily large thermoelectric effect discovered by Seebeck in some materials almost 70 years before and also about the "abnormal" temperature dependence of conductivity discovered by Faraday. Attention was drawn to the discovery made by Ferdinand Brown in 1874: the joint of lead sulphide and metal had a very small resistance when the current was passing along it in one direction and a very large resistance when the current was in the opposite direction. There were dozens of annual publications devoted to the investigation of various semiconductor materials and there were hundreds of papers devoted to the descriptions of various devices made of such materials. Semiconductors became the centre of attention.

However, different investigators could not come to any agreement. The results obtained by them were so different and contradictory that the subject of the investigation was about to be given up. We will speak about this difficult period in the history of semiconductors in detail in the next chapter.

Here we will say that in spite of all the difficulties at the very beginning of the 20th century, a quite correct concept of the nature of semiconductors was formulated. It was due, to a great extent, to the works by I. Köenigsberger, a well-known German electro-chemist.

In 1906, Köenigsberger published his work in which he wrote: "When the temperature of oxides and sulphides is raised, the number of electrons, i.e. of free conducting quanta of electricity is greatly increased until it reaches its limit. Then their behavior reminds that of the metals in which at normal temperature, almost all electrons are free."

THE FIRST REVIEW OF THE PHYSICS OF SEMICONDUCTORS (I. KÖENIGSBERGER)

Eight years later, in 1914, Köenigsberger was the first to publish a review on the properties of semiconductors. Summing up the numerous experiments of the last years, including his own, Köenigsberger introduced the notion of a "class of semiconductors": "Conductors which have the electron conductivity and whose resistance is greatly affected by temperature... will be called semiconductors." Giving a quantitative characteristic of the conductivity-temperature dependence, Köenigsberger gives an absolutely correct determination of one of the most important properties of semiconductors:

$$\sigma = Ae^{-q/T} \tag{7}$$

where σ is the conductivity, T is the temperature, A and q are the constants.

Besides the characteristic conductivity-temperature dependence, the class of semiconductors is characterized according to Köenigsberger by some other important properties: by the magnitude of resistivity within the limits from 10^{-8} up to 10 Ohm/m; by great values of thermo-emf compared with that of the metals and by sensitivity to light.

The word "semiconductor" has remained since the time of Cavendish. But the notion, defined by it, became much clearer. Instead of one rather arbitrary property: the conductivity being neither too good, nor too bad – there were now quite a number of properties, defining the new class of substances. One of the main properties – the conductivity versus temperature dependence of semiconductors – was explained on the basis of the inner processes taking place in crystals. This new class of substances offered new and alluring practical applications.

Chapter 2. *PER ASPERA AD ASTRA* (THROUGH THORNS TO STARS)

> Since everything has been established exactly and firmly,
> And the basis has been prepared by us in its most proper way,
> It is not hard at all to explain the rest.
>
> Lucretius Carus

The previous chapter ended quite optimistically – semiconductors were defined as a separate class of substances. They were investigated most thoroughly and practical applications were found for them. That spoke of progress and success...

HARD TIMES

In 1935, 21 years after the review by Köenigsberger was published, there appeared a new publication on semiconductors by B. Gudden, one of the most famous researchers at that time. Gudden wrote that... there were no semiconductors in the sense in which Köenigsberger had defined them. ... Continuous and thorough investigations show that in this case the relation is rather complicated... Metals, such as graphite, *silicon*, titanium, zirconium, and so on should not be confused with electronic semiconductors...

Reading these words by Gudden we can easily understand why one of the most famous Soviet scientists, academician Abram F. Ioffe wrote in 1956 the words with which we began this part of the book. "Semiconductors is a comparatively young science, it is not yet thirty..." And A. F. Ioffe knew the history of semiconductors quite well.

The reader might think that that was due to the fact that the old use of semiconductors was not valid any longer while there were no new inventions at that time, and as a result, the interest in them faded...

But nothing of the kind! With the development of radio communication at the beginning of the twenties, special devices designed by a wonderful Soviet scientist Oleg V. Losev, crystal semiconductor detectors, were manufactured on a large scale. Besides, in 1926, the American engineer Grondel made a technically perfect rectifier on the base of a semiconductor compound-copper monoxide Cu_2O. Such rectifiers were widely used for charging electric batteries. Furthermore in 1932 on the base of the copper monoxide solid photoelement was designed – a device transforming the energy of light into electric energy.

The reader might think that while the engineers and inventors made experiments, discovered new fields of applying semiconductors and designed new devices, there was confusion in the theory of semiconductors, that no one knew for certain what a semiconductor in fact was. And that prominent physicists might be losing interest in them, and the basic notions remained as undeveloped as during the time of Faraday.

This was by no means true! It was in the thirties that the ideas of quantum mechanics helped physicists to attain good results in explaining the electric, optical and thermal properties of solid states, including semiconductors. In 1931, the English physicist A. Wilson making the best of the ideas of quantum mechanics and using the wonderful work of his immediate predecessors, A. Sommerfeld and F. Bloch, developed a theory of the electro-conductivity of solids. Wilson's theory answered a lot of questions connected with the character of electro-conductivity. A link was found between the structure of the electric shells of atoms and the fact whether the crystal, made of those atoms was a metal, a dielectric, or a semiconductor.

Due to the work of German physicist W. Schottky, English physicist N. Mott and Soviet scientist B. I. Davidov, a theory of metal-semiconductor contact was developed. This theory explained very well the rectifying properties of the semiconductor point detectors.

In 1933, Jacob I. Frenkel explained why the current in the semiconductors was carried not only by the negatively charged electrons, but also by the positively charged holes. Frenkel's theory clarified many electrical properties of semiconductors which until then had not been clear.

Then why did Gudden and some other researchers make such obscure, contradictory and pessimistic conclusions?

Could it be because semiconductors are so sensitive to impurities and because it is so hard to clear the semiconductors from them?

Perhaps yes. Dwelling on the causes of his pessimism, Gudden wrote that in his opinion the conductivity of semiconductors was based on the impurities they contained. If it were possible to clear a semiconductor of the impurities (though Gudden greatly doubted the possibility of it) the "semiconducting" crystal would become an insulator.

But it was of course not only the difficulty of purifying the semiconductors or investigating their intrinsic properties. Abram F. Ioffe as far back as in 1916 showed that there are some ways of cleaning at least certain crystals from impurities and thus allowing the investigation of the conductivity of pure crystals.

Then, why was it so?

Now, 50 years later, it is not difficult to answer this question.

TWO STORIES ABOUT SEMICONDUCTORS

Perhaps we might get a better understanding of Gudden's pessimism if we attend to these two stories. Just two, of the many hundreds and thousands of stories that make up the history of semiconductors.

The first story is about the "oldest" of the semiconductors, which was discovered in 1833 by Faraday – about sulphide silver (Ag_2S).

The Story of Sulphide Silver

Eighteen years later after Faraday had discovered the unusual properties of Ag_2S, there appeared an article by Hittorf, who asserted that there was nothing extraordinary in the conductivity-temperature dependence of Ag_2S. He considered Ag_2S to be nothing else but a solid electrolyte, and declared there was no ground to speak of any special class of substances, mentioned by Faraday.

The properties of electrolyte – the liquids, conducting electricity, had been studied long before Faraday. His teacher, Davy, had investigated a number of such liquids in the years 1800–1807. Faraday also devoted many years of his life to studying electrolytes and established a well-known connection between the quantity of electricity passed through the electrolyte and the mass of the substance released at the electrodes (Faraday's first law for electrolysis). The conductivity of the electrolytes was known to increase with the increase of temperature. Davy and the famous Swedish chemist Berzelius (1779–1848) gave a perfect explanation of this phenomenon. The current in the electrolytes is transferred by both positively and negatively charged particles (ions), moving to the opposite electrodes. The higher the temperature, the easier the electrolyte decomposes into positive and negative ions, the higher then is the concentration of ions and consequently the greater is the value of conductivity.

Moreover, it was known that the electrolytes were not necessarily liquids. Davy was the first to obtain the element potassium when passing the electric current through the solid electrolyte-caustic potash (KOH).

By the time Hittorf published his article, Faraday's authority was indisputable. In order to accuse Michael Faraday of his not having noticed the silver released on the electrodes, Hittorf must have had very good grounds. And he really had them; in his experiments he had silver released on the electrodes.

A few dozens of years later the researchers had at their disposal a new and very effective method of investigating electroconductivity – the Hall effect. It allowed them to determine the mobility of the charge carriers. The mobility of ions is known to be thousands of times smaller than that of electrons or holes. The higher the mobility of the carriers, the easier it is to see the Hall effect. Why so, we will learn later in the third part of this book where we will discuss this effect in detail. During the first scores of years after the discovery of the Hall effect, the sensibility of the method was so low, that it was impossible to detect the motion of ions by means of this method. The Hall effect was not observed in the electrolytes where the current was carried by ions. But it was possible to see this effect in substances where the current was conducted by electrons.

In 1902, Straints, having investigated Ag_2S came to the conclusion that Faraday was right and Hittorf was wrong – the conductors of electricity in Ag_2S were electrons.

But in 1920, Tuband and his assistants not only proved the release of silver at the electrode when the current was passing through Ag_2S, but they also measured its quantity, thus confirming Faraday's First Law. They found that the current was transferred by the ions of silver; the amount of silver released at the electrodes was proportional to the charge passing through them, the error being approximately 1% and the coefficient of the proportionality exactly corresponded to Faraday's law. So, it seemed to be clear that Hittorf was right, while Faraday and Straints were both wrong.

But the results obtained by Tuband threatened to shake the laws of electro-chemistry because if the conductivity of Ag_2S was due to the current of the ions of silver, then those ions should have the mobility 1000 times greater than that which followed from the most optimistic theoretical estimation.

The problem was solved by the time of its hundredth anniversary. In

1933, K. Vagner discovered the source of error in Tuband's experiments. The release of sulphur and silver on the electrodes took place not as a result of electrolysis but as a result of the accessory reaction on the electrodes. It was possible to create certain conditions under which that reaction would not take place. In that case no silver would be released on the electrodes and still it would not affect the conductivity of silver sulphide. So, it was the final proof of the fact that Michael Faraday was right.

But it took a whole century to make it clear.

The Story of Copper Monoxide

Copper monoxide (Cu_2O) was one of the first semiconductors studied most thoroughly. Being widely used in rectifiers and photoelements, this material aroused much interest.

Figure 11 shows the dependence of the conductivity of the copper monoxide on temperature, obtained by three different research groups. (Pay attention to the scale on the Y-axis. One grade displacement corresponds to the tenfold change of conductivity.)

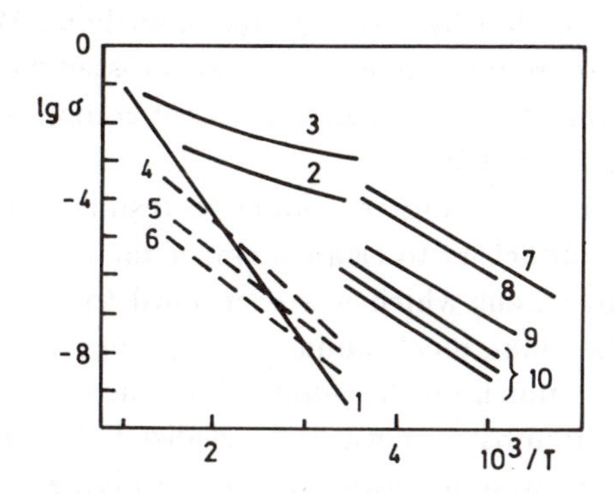

Fig. 11. The conductivity of copper monoxide (Cu_2O) versus temperature (σ Ohm^{-1} cm^{-1}). Curves 1–3 by V. Zhuse and B. Kurchatov; 4–6 by Leblan and Zaks; 7–10 by E. Engelgard.

The Soviet physicists, V. P. Zhuse and B. V. Kurchatov measured the conductivity of the compound Cu_2O cleaned to such an extent that the excess oxygen, which is usually present there, was not detected. Curve 1 shows the conductivity-temperature dependence for such pure specimens. If the specimens contained oxygen impurity at concentration of 0.06% or 0.1% then the dependence σ(T) was determined by curves 2 and 3 respectively. But the results obtained by Leblan and Zaks for the purified Cu_2O were quite different (curves 4, 5, and 6). At 700 K (1000/T = 1, 4) as it is seen from the figure, the conductivity of specimens was by their data quite different and was much smaller than the value established by the Soviet physicists. The figure shows that for the specimen of Leblan and Zaks which had the smallest conductivity (curve 6), the value σ was 200 times smaller than the value found by Zhuse and Kurchatov! Needless to say, the difference is remarkable.

Curves 7–10 were obtained by E. Engelgard, another scientist investigating the conductivity of Cu_2O.

Trying "to freeze" the excess oxygen, Engelgard made all his measurements at a comparatively low temperature. He hoped that under such conditions he would obtain the most reliable results which could be easily repeated. As it is seen from the figure, he obtained the values of conductivity which differed for various specimens more than 1000 times!

Figure 11 is taken from Gudden's review, by which we began this chapter, although it is rather simplified here. In the original the results obtained were not given by three but by seven groups of authors. Without Gudden's commentaries, the figure looks like a rectangle crossed with lines at various angles and at all levels, So, it is clear that "most complicated relations take place in this case...", isn't it?

In order to get reliable and reproductive results it is necessary when working with semiconductors to clean them, reducing the concentration of impurity to such an extent which was even hard to imagine at that time. Never before had the main, most fundamental properties of the investigated material depended so much on the infinitesimal, negligibly small quantities of impurities. This circumstance was to be realized. It was also necessary to understand that there were no other ways for studying semiconductors but by elaborating new perfect methods of purifying them. Physicists and engineers were to unite their efforts and to elaborate such methods of physics and chemistry and such devices which would provide the opportunity of spotting and investigating the negligibly small quantities ("traces") of impurities and designing the necessary devices for measurement.

FIRST QUEEN OF RADIOELECTRONICS

In the thirties of this century, scientists were not only interested in semiconductors but also in other branches of physics which were developing very rapidly. And so the vacuum valve (tube), the glorious invention of the Americans, Thomas Edison and Li de Forest* became the main element of the first "revolution in electronics".

In the thirties of this century, the physical principles of the point semiconductor detector could at last be understood. But by the irony of fate the detectors by themselves were then of no interest to anyone.

Vacuum valves provided a much wider range and a much better quality of radio reception than the detector designed by Losev. The vacuum valves were provided with the most reliable amplifiers, durable and stable. They could be sunk to the bottom of the ocean together with the transatlantic telephone cables and could be lifted into the sky with airplane radiosets. The vacuum valves served as elements of generators, transformers, detectors and amplifiers. In 1927, the American company, Bell Telephone demonstrated the first industrial TV tube receiver. It seemed that for years to come the vacuum valves would ensure the progress of the newly born science, radioelectronics, so essential for physicists and communication men, for medical and military men, for navigators and criminologists and for people of many professions. Semiconductor devices were not to be compared with them. It was a real revolution in electronics.

LONG LIVE ENTHUSIASTS!

But... and this is observed in the history of any science: no matter what turn circumstances take, no matter how much out of date this or that direction in science might seem, there are always enthusiasts who continue their investigation in the branch of science they have chosen, in spite of the difficulties and complications they come across.

The same happened in the case of semiconductors. Practically in every developed country there were enthusiasts who went on investigating those materials and searched for new applications. In Russia, this work was headed by A. F. Ioffe, a remarkable physicist and a brilliant organizer.

---

*Li de Forest was not only a remarkable inventor but also a gifted physicist, the pupil of Willard Gibbs, one of the greatest physicists of the last century.

Abram F. Ioffe
(1880–1960)

Abram F. Ioffe was born on October 29, 1880 in the little town of Romny in the Ukraine. On leaving the Romny technical high school, Abram Ioffe entered the Petersburg Polytechnical Institute. On graduating from the Institute he was offered a very good job with a high salary. But Ioffe declined that proposal and left Petersburg for Munich, Germany. He wanted to specialize in experimental physics and began working at the laboratory of Wilhelm Conrad Röntgen.

W. Röntgen, who discovered the radiation named after him, was one of the best physicists of his time. That laboratory was a very good school for the young scientist from Russia. It was a school of experimentation, of devotion to science and of utmost scrupulousness in scientific research.

In Munich, Ioffe made his first experimental investigations and in 1905 he defended his Ph. D. thesis which was marked with a special reward. He was offered the post of a professor, but he chose instead to return to his motherland.

In 1916, Ioffe organized in Petrograd a seminar on modern physics. The seminar was attended by P. L. Kapitsa and N. N. Semyonov – both of whom later became Nobel laureates – by P. I. Lukorsky, J. I. Frenkel, J. G. Dorfman and other well-known scientists.

In 1918, through Ioffe's initiative, the Physical-Technical Institute was established. It soon became famous all over the country.

A. F. Ioffe was one of the first to forsee the wonderful prospect of the scientific and technical aspects of semiconductors.

Under the guidance of Ioffe in 1928, the Physical-Technical Institute began elaborating thermo- and photoelectric devices and semiconductor refrigerators.

In 1933, Ioffe published the first book on semiconductors in Russian.

Abram F. Ioffe had cordial relations with many remarkable scientists all over the world: Röntgen, Ehrenfest, Einstein, Bohr, Debye, Lorentz. He had written an interesting book about them, *Meetings with Physicists.* A. F. Ioffe died on the 14th of October, 1960, a fortnight before his 80th birthday.

It was due to the work of enthusiasts such as Ioffe that only ten years after the pessimistic forecast of Gudden, semiconductors came to be the centre of attention throughout the world.

THE QUEEN IS GETTING TOO OLD

The thirties ended with a catastrophe that affected all mankind. The Second World War brought about enormous numbers of victims and massive destructions. The war showed that the most abstract scientific researches could result in inventing new arms and introducing fundamental changes in industry. It became obvious that though the role of an individual in forming scientific ideas and making important discoveries was very great, it was also essential to organize and coordinate the work of the large collective bodies of researchers, designers and engineers.

Meanwhile, with time the necessity to acquire more knowledge concerning semiconductors grew more urgent. At the end of the thirties the development of radiolocation made engineers recall the semiconductor detector. It turned out that the point detectors made of germanium and of silicon could transform the microwave signals, for which the tubes (valves) of that time were absolutely useless. Semiconducting photoelements, thermobatteries, selenium and copper monoxide rectifiers found wide applications in techniques.

But the main thing was that "the queen of radioelectronics" – the vacuum valve could not help in solving the problems which were becoming always more difficult. It took much time for the tubes, many seconds and even minutes, to be heated, and it was the time which could not be afforded under certain circumstances – when minutes might decide the fate of an airplane, which there was no communication with, or of a warship whose radar was out of order, or of the supersonic locator amplifier which was to warn in case of the appearance of the enemy submarine.

The lifetime of the vacuum tube is limited. The incandescent cathode of the valve, emitting electrons, gradually gets spoiled and the valve is put out of action. If the electronic scheme contains only several tubes, there is no problem, but the consequences may be quite grave if the device comprises thousands or tens of thousands of elements and the valve cannot be replaced immediately, for instance in the amplifier of the intercontinental telephone cable.

The urgency of creating a principally new element with an unlimited lifetime which would not take time to be heated, and would not be easily affected by any external action, became quite obvious.

The most prescient specialists began thinking of a semiconductor triod – a transistor.

In fact, attempts to design such a semiconductor device – a transistor – were made as far back as in 1926. Various semiconductor materials were tested, as well as various configurations of electrodes. Many countries gave patents on hypothetical solid amplifiers, but during the twenty years that passed not a single device had been created.

Success was attained only after the best physicists, future Nobel prize laureates, and also a lot of other specialists: metallurgists, chemists, technologists and engineers took part in that work. It was due to their joint and coordinated efforts that success was achieved.

THE KING HAS DIED – LONG LIVE THE KING!

Bell Telephone Company was the first to demonstrate a solid amplifier – a point transistor made by the engineers of the company – Bardeen, Brattain and Shockley (Fig. 12) a year before.

In Fig. 13 we see a point transistor which was demonstrated to specialists in 1948. Pay particular attention to the touching attachment of human imagination to traditional forms: the exterior of the first transistor repeats that of the vacuum valve. Thus the forms of the first automobiles repeated those of coaches and carriages. That went on for 15 to 20 years, though it took the designers of transistors much less time to change the form of the transistor. Two years later, transistors were produced in special cases which did not resemble at all any vacuum tubes.

Hundreds and thousands of specialists of various professions in industrially developed countries began their work on improving the transistor. Metallurgists, physicists and chemists joined their efforts in trying to find a way of creating new materials. Technologists elaborated new methods of purifying materials so as to create perfect monocrystals. As soon as physicists obtained pure materials with reproducible parameters, they prompted new ways of improving transistors and furthermore they discovered new physical effects. Some of those effects might be of help in designing new devices. Engineers, physicists and designers who had been elaborating these devices made new demands on the materials and to the degree of their purification. Metallurgists, chemists and technologists...

Fig. 12. The creators of the transistor, future Nobel prize winners in physics, W. Shockley (left), W. Brattain (center), and J. Bardeen (right) (from *Electronics*, 1948). They were awarded the Nobel prize in 1956 for their research in the field of semiconductors and for creating the transistor.

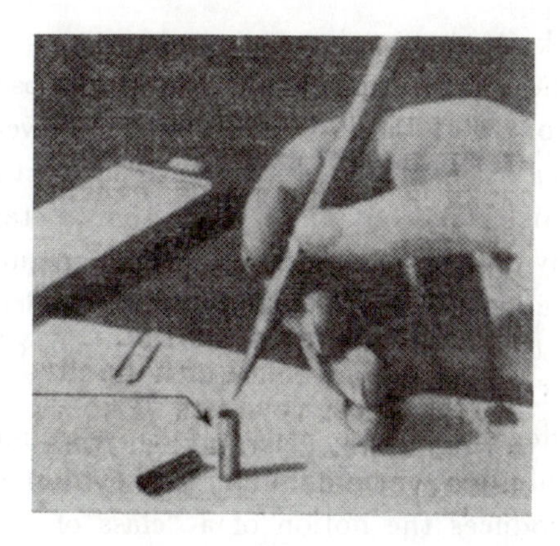

Fig. 13. The first germanium transistor (indicated by the point of a pencil at the bottom of the figure). For the sake of comparison there is also a valve on a sheet of paper – the smallest of the valves was manufactured in 1948 (from *Electronics*, 1948).

The magic "merry-go-round" that was set in motion with the invention of a transistor was moving faster and faster. Nowadays it is rotating at fantastic speeds.

Hundreds of semiconductor materials have been investigated and a lot of new semiconductors are synthesized every year. The purity level of the semiconductor materials as well as the methods of control have become quite perfect. Sometimes specialists manage to detect one impurity atom for ten thousand billion intrinsic atoms. Technologists elaborated the method of producing monocrystals of enormous sizes. They can grow silicon monocrystals whose volume is 6000 cm^3. They have invented ways of growing atomic layers, under the control of computers; layer after layer, and even atom after atom. There are now thousands of semiconductor devices based on the effects which have been discovered by physicists. Billions of semiconductor devices and millions of integral circuits are annually manufactured throughout the world, each circuit containing tens and hundreds of thousands of separate diodes and transistors.

Semiconductors and semiconductor devices are now part and parcel of modern civilization and to a great extent determine its development.

A FAREWELL GLANCE ON WHAT WE HAVE COVERED

Let us look back at what we have covered.

The sixties of the 18th century saw the works by Joseph Priestley, Jambatista Beccaria, John Canton and, finally, Henry Cavendish. They maintained that there is a group of substances which conduct the electric current much better than dielectrics but much worse than metals.

In 1833, Faraday discovers that when the temperature is changed, some substances display a most unusual temperature dependence of conductivity. Faraday suggested that this property must be a sign of the existence of a certain new class of substances.

The first decade of the 20th century: On summing up the numerous results of hundreds of works by scientists from different countries, Köenigsberger introduces the notion of a "class of semiconductors" and names the main distinctive features of this class.

The twenties and thirties of the 20th century: the fundamentals of the physics of semiconductors are laid down by the works of A. Wilson, V. Schottky, N. Mott, A. F. Ioffe, V. Davydov. J. Frenkel, and of many others.

The year 1947: Bardeen, Brattain and Shockley create the first semiconductor triode – the transistor. This discovery, answering the most urgent and vital practical needs, attracted the attention of many scientists, technologists, and engineers. United into large collective bodies, they combined efforts, using for their research the best up-to-date equipment and the most perfect methods of investigation. Since that time the progress in studying semiconductors has to a great extent determined the development of the most important branches of industry, communication, military science, domestic facilities and the researches of the cosmic space...

Per aspera ad astra!

Part II. THE SECRET OF THE GOLDEN MEAN
The main properties of semiconductors

The mean is the point nearest to wisdom. Not to reach it is the same as to get over it.

Confucius

On getting acquainted with the basic properties of semiconductors and with the story of their discovery and investigation we came to realize that "semiconductors" denote in fact not just those substances whose conductivity is worse than that of metals and better than that of dielectrics, but a notion much wider and deeper.

In this part of the book, while getting better acquainted with the main heroes of our story – the electron and the hole – we will continue investigating the physical properties of semiconductors.

Chapter 3. THE PRINCESS AND THE PEA (IMPURITY SEMICONDUCTORS)

Now everyone understood she was a real princess, for she lay on a dozen mattresses and a dozen feather quilts, and still she felt the pea beneath them. Only a real princess could be so sensitive.

Hans Christian Andersen

When discussing in the Introduction the properties of intrinsic semiconductors, we could see that many of them were very good insulators even at room temperature. Indeed, the concentration of intrinsic electrons in gallium phosphide is 10^{23} times smaller than in metals (see Table 1). In silicon carbide it is as much as 10^{30} times smaller! Even the intrinsic gallium arsenide is a very good insulator.

Whether such substances can be used to make semiconductor devices – depends on the properties and the quantity of the impurities. So the impurity in a semiconductor may sometimes be helpful. The great majority of semiconductor materials contain a certain controlled amount of impurities which provide the necessary value of the conductivity.*

Let us see how the density of free carriers in a crystal is affected by impurities. We will begin with the simplest example.

DONOR IMPURITY

Let us assume that an arsenic atom has somehow got into a silicon crystal and occupied a place in one of the sites of the crystal lattice, substituting its lawful host – the silicon atom. Now let us look at Fig. 3. Let us suppose that the As atom is in site 13 of the crystal lattice of silicon. The silicon atom has four valent electrons; the arsenic atoms have five valent electrons. The four valent electrons of arsenic atoms make bonds with the neighboring silicon atoms (atoms 8, 12, 14 and 18). And what about the fifth electron?... The fifth electron will be held by the As atom, but much more weakly than the other four electrons, which are very tightly bound in their electron orbits, determined by the structure of the silicon crystal. The energy ΔE, necessary to break the bond of the fifth electron with the arsenic atom and to transform it into a free electron, is much smaller than the energy E_g which is necceassry to break the bond between the silicon atoms and form an electron-hole pair.

The impurity whose atoms give away their electrons easily is called a *donor impurity*. The Latin word "donare" from which the word "donor" comes, means "to endow", "to give".

---

*That is one of the reasons why semiconductors and dielectrics cannot be distinguished by the value E_g. Diamond, for instance, has $E_g \approx 5.6$ eV. Using Eq. (6) we can calculate that to find a single free electron in a diamond at room temperature, it is necessary for the diamond to be as large as the Earth. Nevertheless in scientific literature devoted to semiconductors, we may come across the expression: "a semiconductor diamond". It appeared about thirty years ago when blue diamonds were found in the mines of South Africa. For some unknown reason, in the process of their natural birth they came to get some boron impurity and due to that they acquired an impurity conductivity.

Let N_d atoms of a donor, say of arsenic, be introduced into every cubic centimeter of a crystal. Now let us first consider the simplest situation – the temperature of the crystal is $T = 0$ K. It is clear that in this case the crystal remains an ideal dielectric: though it takes a very small amount of energy to break away the fifth electron from the arsenic atom, the temperature being equal to absolute zero, there is no energy of thermal oscillations there whatsoever.

If the temperature of the crystal is $T > 0$, then the equilibrium density of the impurity electrons n_d is determined by the expression, analogical to Eq. (6)

$$n_d = (A \cdot N_d)^{1/2} e^{-\frac{\Delta E}{kT}} \ . \tag{8}$$

Instead of a large value E_g, we have in the exponent a much smaller value ΔE. The ionization energy ΔE (which is sometimes called the activation energy of impurity) of arsenic in silicon is very small, it is equal to 0.05 eV, which is twenty times smaller than the energy E_g necessary to create an electron-hole pair in silicon. According to Eq. (8) (and according to common sense) it means that the ionization of the arsenic atoms, i.e. the detachment of the extra fifth electron, will take place at a temperature much lower than that of the generation of electron-hole pairs.

It would be useful to compare the densities of the intrinsic electrons (n_i) and the impurity electrons (n_d) at different temperatures. To do it we must first of all know the impurity density N_d in Eq. (8). This value can change within a very wide range.

The lowest possible value of N_d depends on the level of technology and on our skill in purifying the semiconductor. For silicon carbide, for example, the lowest value of impurity density is $10^{16} - 10^{17}$ cm^{-3}. For gallium arsenide this value is $10^{13} - 10^{14}$ cm^{-3}. Technologists have acquired great skill in purifying silicon. This is of utmost importance as most of the semiconductor devices are now manufactured from silicon. The level of "uncontrolled" impurity in silicon ranges from $10^{10} - 10^{11}$ cm^{-3}.

Let us now consider silicon into which a donor impurity (As) has been introduced. The density of the impurity atoms is $N_d = 10^{15}$ cm^{-3}.

This implies that there are 10 000 000 Si atoms per single As atom. From a chemist's point of view there is no arsenic in the silicon at all. But the value $N_d \approx 10^{15}$ cm^{-3} is very typical for many cases of practical importance.

In silicon $A \approx B \approx 10^{19}$ cm^{-3}. Now by Eqs. (6) and (8) it is quite easy to calculate that with the temperature at 10 K, the density of the impurity electrons would make $n_d \approx 2.5\,10^4$ cm^{-3}, while the density of the intrinsic electrons would be so small that even if a crystal were as big as the Galaxy there would not be a single intrinsic carrier in it.

At a temperature of 50 K the value n_d is equal to $\sim 3\,10^{14}$ cm^{-3}, while the concentration of the intrinsic electrons is still practically equal to zero in any real crystal. Note that at this temperature about one third of the impurity atoms prove to be ionized.

As for room temperature (300 K), it would be wrong to calculate the concentration of the impurity electrons by means of Eq. (8) since it is not exact but just an approximation. It can be used only in those cases where the value n_d, calculated by it, is much smaller than the concentration of the impurity atoms N_d, i.e. when $kT \ll \Delta E$.

But if the temperature is so high that $kT \sim \Delta E$ (or moreover if $kT > \Delta E$), then all of the impurity atoms prove to be ionized and the concentration of the impurity electrons is just equal to N_d. This phenomenon is called the "impurity saturation (exhaustion)" – the term seems rather reasonable. When the temperature is high enough, all the impurity atoms give away their "extra" electrons and then the source of the electrons is "exhausted". No further rise of the temperature of the crystal will increase the concentration of the impurity electrons.

In Si, room temperature (300 K) corresponds to the impurity saturation region. So in our example when $T = 300$ K, $n_d = N_d = 10^{15}$ cm^{-3}. While the concentration of the intrinsic electrons n_i at 300 K is equal to only $\sim 10^{10}$ cm^{-3} (see Table 1), which is 100 000 times smaller.

Yes, the Princess on the pea is a pachydermatous hippopotamus in comparison with the most ordinary silicon! Her Majesty could feel a pea through a dozen mattresses and a dozen feather quilts. While after arsenide has been introduced into silicon (1 atom of arsenide per ten million atoms of silicon) the concentration of electrons into silicon crystal becomes 100 000 times greater.

But what about the intrinsic electrons and holes? Didn't they play any role at all? That would be a hasty conclusion. Let us heat a crystal up to $T = 1000$ K. The concentration of the impurity electrons will still be 10^{15} cm^{-3}, while the concentration of the intrinsic electrons, calculated in Eq. (6), will be equal to $n_i \approx 1.5\,10^{16}$ cm^{-3}. The number of intrinsic electrons in Si at 1000 K will prove to be 15 times greater than that of the impurity electrons.

THE TEMPERATURE DEPENDENCE OF THE ELECTRON DENSITY

Look at Fig. 14. The curve, shown there, gives a short summary of what we have learned about the properties of both – intrinsic and impurity semiconductors. The curve represents a typical temperature dependence of the equilibrium density of free electrons in a semiconductor. Though the coordinates chosen to illustrate this dependence are somewhat unusual:

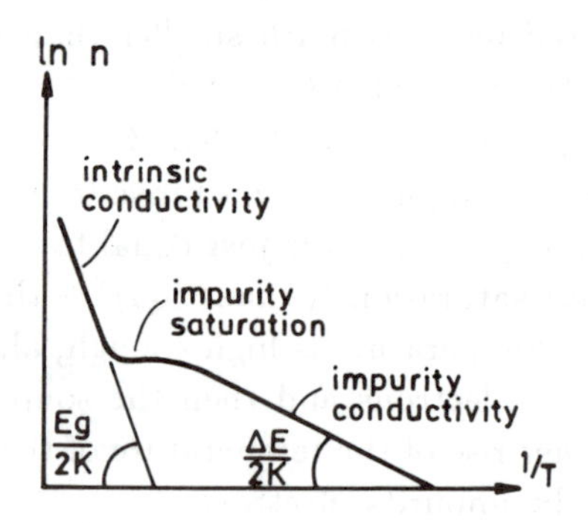

Fig. 14. Electron density logarithm versus $1/T$.

it's not the density n, but the value of its natural logarithm $\ln n$ which is plotted on the y-coordinate. And it's not temperature, but its inverse value $1/T$ is plotted on the abscissa. So, it makes the "reading" of the graph a bit difficult, especially since a movement to the right on the abscissa corresponds to a decrease of temperature rather than to its increase. (The value $1/T$ rises, and the value T falls.) We will explain now why such coordinates have been chosen. But let us first "read" the graph. We will begin from the right from the low temperature region, where the value $1/T$ will be great, and we will advance on the abscissa from the right to the left. As we have seen at low temperatures the electron density in a semiconductor is determined by the density of impurity centers. The electron density increases with the rise of temperature, and at this section of the curve, the dependence $n(T)$ is determined by Eq. (8). At a certain temperature this dependence is saturated. It is the impurity saturation region. All the impurity atoms have already been ionized, but the intrinsic density is still much smaller than that of the impurity. Finally, in the region of a still higher temperature there is an abrupt increase of the density with a further rise of the temperature. It is the region of intrinsic conductivity where the $n(T)$ dependence is expressed by Eq. (6).

And now a few words about the choice of the coordinates in Fig. 14. Let us take the logarithms of Eq. (6)

$$\ln n_i = \frac{1}{2}\ln(AB) - \frac{E_g}{2k}\frac{1}{T} \tag{9}$$

and we will denote $x = 1/T$; $y = \ln n_i$; $a = \frac{1}{2}\ln(AB)$, $b = E_g/2k$, Eq. (9) will look familiar: $y = a - bx$ – the equation of a straight line. So, if we plot $\ln n_i$ on the y-coordinate, and $1/T$ on the abscissa until Eq. (6) is valid, then the dependence $\ln n_i(1/T)$ must be a straight line. And, what is especially important, the slope of the curve of this straight line $b = E_g/2k$ is just directly proportional to the most important parameter of the semiconductor – E_g. In the early works on semiconductors this way of defining E_g was used very often. It is sometimes used nowadays too, when studying the properties of new semiconductors.

If the same operation of taking the logarithms is repeated with Eq. (8), it is easy to see that the slope of the straight line $\ln n(1/T)$ is proportional to the value ΔE. Hence, having plotted the electron density-temperature dependence in the coordinate $\ln n(1/T)$ in the impurity conduction region, we can easily determine the amount of the impurity activation energy ΔE.

HOLES IN THE ELECTRONIC SEMICONDUCTOR

A semiconductor into which some donor impurity has been introduced is called an electronic semiconductor, or a semiconductor of n-type.

The term "electronic semiconductor" is simple to understand. No other term could be more natural for a semiconductor in which almost all of the free charge carriers are electrons. As for the semiconductor of n-type, the letter "n" comes from the word "negative", showing that the semiconductor contains many negatively charged particles – electrons.

But our talk on the semiconductors of n-type would be incomplete if we forgot to speak of the holes in semiconductors of n-type. Holes in this case are called the *minority carriers*. (One can easily guess that electrons in this case are called the *majority carriers*.) First of all it should be mentioned that when the donor atom gets ionized and yields an electron, no hole is formed. The arsenic atom in Si lattice, for instance, having yielded the fifth valent electron remains bound with the adjacent atoms of silicon by four "valid" bonds. There is no empty place there to which electrons from the adjacent orbits could be displaced. Therefore no hole is formed.

On the face of it, it might seem that the electronic semiconductor must then have as many holes as the absolutely pure intrinsic semiconductor. But that view would be wrong. It is true that at a given temperature T the number of broken electron bonds and electron-hole pairs, appearing every second in an electronic semiconductor, is exactly equal to those in an intrinsic semiconductor. But the holes perish much more often, because there are a lot more free electrons in the electronic semiconductor than in the intrinsic one. That means that a hole will meet with an electron much more often and each time it will result in a recombination and the disappearance of the hole. As a result of this there are fewer holes in the n-type semiconductor than in an intrinsic semiconductor. There is a very simple equation connecting the equilibrium density of holes p and the electrons n in a semiconductor

$$p \cdot n = n_i^2(T) \ . \tag{10}$$

The greater the number of electrons, the smaller the number of holes and vice versa.

The quantity n_i in Eq. (10) represents the value of the intrinsic carrier density at a given temperature T and is defined by Eq. (6).

By using Eqs. (10) and (6) it is easy to calculate that in n-type silicon, say, at room temperature, and with the electronic density of 10^{15} cm^{-3}, the hole density will be $\sim 10^5$ cm^{-3}, i.e. ten billion times smaller than that of the electrons! It is clear now which of them are the majority carriers.

We can't say yet that we're well acquainted with impurity semiconductors. We have just discussed the electronic semiconductors into which some donor impurity has been introduced. But to do justice and also for practical reasons, we have to consider another situation, namely when the impurity does not yield its electrons (as the donor impurity does) but on the contrary takes them up from the nearby atoms.

ACCEPTOR IMPURITY

Figure 15 shows a crystal lattice of silicon in which one of the sites is occupied by the boron impurity atom. Boron (B) is trivalent – there are three electrons on its outer electron shell. It is one electron short of making a complete bond with the neighboring silicon atoms.

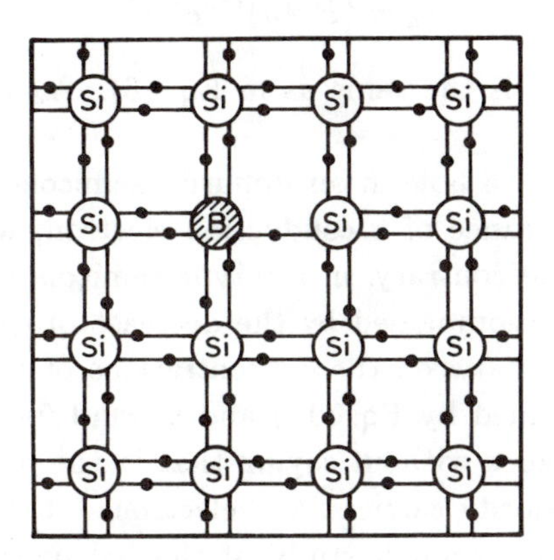

Fig. 15. The acceptor atom (boron) in the silicon lattice.

Compare the picture of the electron bonds of the B atom in Fig. 15 with the configuration of the electron bonds around atom 19 in Fig. 3. The situations are quite similar, aren't they? Both, the silicon atom 19 in Fig. 3 and the B atom in Fig. 15 are short of one electron. But there is also a great difference between them. All the silicon atoms are quite identical, and the empty link, the hole, which now belongs to atom 19 may at any moment approach atom 14, then atom 9 and so on.

It does not take any energy for the hole to travel in the crystal. But a boron atom is a stranger in the silicon lattice. To make the electron of the neighboring silicon atom go over to the boron it is necessary to expend energy ΔE. This energy, the activation energy, is not large (for B atoms in Si it is only 0.045 eV), and yet it is not zero, as in the case of a hole in Fig. 3. No hole will be formed in the crystal until the "energy barrier" ΔE is overcome, no matter how small it is.

Let us assume that either the lattice vibrations or a quantum of light have supplied the necessary energy, and that the electron from the neighboring Si atom has come over to boron. Now the situation will be identical to that shown in Fig. 3. There is an empty link in the lattice, i.e. a hole – a free carrier of positive charge. So, acceptor impurities create holes – free carriers in a semiconductor crystal. If the temperature of the crystals is $T > 0$, then the equilibrium density of the impurity holes p_a can be found from the expression analogous to Eq. (8).

$$p_a = (BNa)^{1/2} e^{-\frac{\Delta E}{kT}} \tag{11}$$

The value B here is the same as in Eq. (6). N_a is the density of the acceptors.

The appearance of a hole in an impurity semiconductor is not accompanied by the appearance of a conduction electron, which is clearly seen from Fig. 15. On the contrary, in a p-type semiconductor the increase of the hole density is accompanied by the decrease of electron density. The quantitative relation between the concentrations of the majority and minority carriers is defined by Eq. (10) and is valid for both n- and p-type semiconductors. It goes without saying that in the p-type semiconductor, holes will be the majority carriers, and electrons – the minority carriers.

We have made a thorough study of the behavior of *shallow impurities.* (The impurity is called "shallow" if its ionization energy ΔE is much smaller than E_g). The attention we have paid to donor and acceptor shallow impurities is quite justified.

To manufacture a semiconductor device it is necessary for a certain, precisely calculated amount of shallow impurities to be introduced into a thoroughly purified semiconductor material. The concentration of impurity electrons (or holes) as a rule determines the parameters of the devices.

The role of *deep impurities* in the physics of semiconductors and semiconductor devices is also very great. (By deep impurities we mean impurities whose ionization energy is comparatively great and is of an order comparable to E_g.) We will discuss this kind of impurity in the next chapter.

To conclude this chapter we will say a few words on the methods of growing the semiconductor crystals and purifying them to the extent that they can be used in semiconductor device production.

CRYSTAL GROWTH AND PURIFICATION OF SEMICONDUCTORS

The Czochralsky Method

The Czochralsky method consists in pulling the monocrystal from the semiconductor melt.

The main principles of this method are indicated in Fig. 16. A small monocrystal fixed on a holder is plunged into the melt of a semiconductor. This is the so-called "seed". The seed is rotated and is slowly lifted over the surface of the melt. The surface tension forces make some quantity of the liquid follow the seed. Over the surface of the melt the liquid is crystallized and ... becomes a seed for the next portion of the liquid semiconductor. The monocrystal on the holder is gradually growing, getting quite massive.

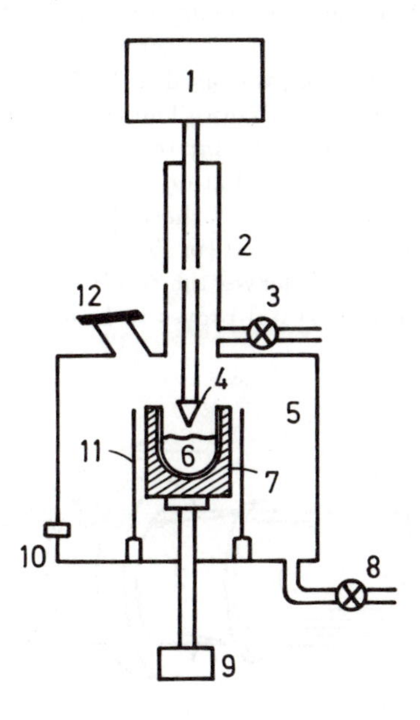

Fig. 16. Schematic diagram of an installation for growing semiconductor monocrystals by the Czochralsky method: 1 – pull rod; 2 – the outer chamber; 3 – inert gas entrance; 4 – the seed-holder and the seed; 5 – high temperature chamber; 6 – the melt; 7 – a crucible, made of quartz (SiO_2) or of silicon nitride (Si_2N_4); 8 – gas outlet; 9 – a set for lifting and rotating the crucible; 10 – a temperature sensor; 11 – the heaters; 12 – the peep-hole.

To protect the growing crystal from any pollution, the growing is either done in vacuum, or in the atmosphere of an inert gas. The holder is rotated and moved up and down by means of magnets attached to the outside of the installment. The melting is done by microwave heating by means of a radio-frequency coil which is also outside the installment.

Huge silicon monocrystals, used to manufacture powerful semiconductor devices, are grown by the Czochralsky method and also by the method of floating zone melting. Their length may reach ~ 50 cm, and their diameter ~ 20 cm.

In the process of growing a crystal by the Czochralsky method the semiconductor can be doped by adding to the melt donor impurities (for obtaining an n-type semiconductor), or acceptor impurities (for obtaining a p-type semiconductor).

Floating Zone Melting

One of the ways of obtaining pure semiconductor crystals is that of the floating zone melting method.

For the first time zone melting was used in 1952 to purify germanium. It was done by the American scientist B. Pfann. The main idea of this method is shown in Fig. 17. The rod of a comparatively impure semiconductor is placed in the field of a powerful source of microwave radiation, created by an inductive coil. When the coil is switched on, a small part of the rod, which is inside the coil, melts. The coil is moving slowly along the rod (in Fig. 17 it is from the left to the right). When the melted part of the rod leaves the coil it is cooled (it crystallizes). This newly crystallized part of the crystal proves to have much fewer impurities than the initial material.

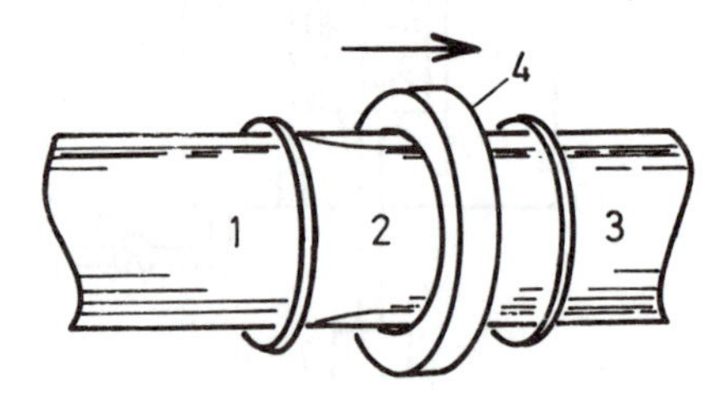

Fig. 17. Schematic diagram of zone melting: 1 – a section of a crystal purified by passing through the zone; 2 – the melt (the zone); 3 – a section of a starting semiconductor; 4 – a loop of an inductor.

The method of zone melting is based on the phenomenon of impurity segregation which is well-known in chemistry. It is due to this phenomenon, that ice which is crystallized from salty sea water is practically saltless. Polar explorers are well informed of the phenomenon of segregation. When starting their Arctic voyages, they do not take any fresh water with them. They are aware of the fact that fresh water is available there – it is just necessary to melt a piece of ice (formed from the salt water of the Arctic Ocean!) and the travellers have as much water as they like at their disposal.

The segregation is based on the fact that when the melt is cooling, forming a crystal lattice, it is energetically more expedient to arrange the intrinsic atoms rather than the impurity ones in this new lattice.

Let us imagine a pool of salt water which freezes as the temperature is lowered. As a rule, ice is first formed on the edge of the pool. If you break off a thin transparent plate of ice and try to taste it, licking it with the tip of your tongue, you can be sure it will be quite saltless. But where is the salt? When the ice crystal lattice was formed, salt was displaced from the solid phase (ice) to the nearby liquid phase (water). It proved to be much more expedient for the new lattice to extract its own "intrinsic" atoms from the melt, leaving the impurity ions of Na and Cl in the liquid phase.

The relation of the impurity concentration in the solid phase to the same impurity concentration in the nearby liquid phase is called the segregation factor K_s. The smaller the value of K_s, the more effective the purification of the crystal in the process of crystallization.

The floating zone melting method is used most often to purify silicon – the most important semiconductor material of modern semiconductor electronics. The following table gives the segregation factors for some of the most important impurities contained in silicon.

Impurity	Cu	Zn	B	Al	Ge	P	Sn	As
K_S	0.0045	10^{-5}	0.8	0.002	0.33	0.35	0.016	0.3

It is seen from the table that copper, zinc and aluminium are removed from the growing monocrystal most effectively. During one pass of the zone along the silicon crystal, the concentration of Cu decreases $1/K_s = 220$ times. Of Al – 500 times and of Zn – 10^5 times. It is much more difficult to remove the Ge, P, Sn and As impurities. Therefore it is sometimes necessary to have the zone pass up to 50 times along the rod to purify it, so as to produce monocrystals which would answer modern requirements.

But the boron (B) impurity proves to be practically invulnerable for zone melting. It is necessary to use special methods to get rid of it. Before silicon is subjected to zone purification, it is melted and placed in the atmosphere of humid hydrogen, which forms a fugitive compound with boron – BH_3. Boron disappears, and all the other impurities are sure to be removed in the process of zone melting.

The initial material to be purified by the zone melting method is the so-called technical silicon, it is obtained from natural silicon oxides, sand or quartz, by reduction in the flame of the electric arc.

The floating zone melting method allows the purifying of silicon to the concentration of $\sim 10^{12}$ impurity atoms per 1 cm^3, i.e. to the level corresponding to 1 impurity atom for 10 billion Si atoms. And the question is not of growing some unique crystals. Such super-pure crystals are produced on an industrial scale, with an annual output of tens of tons of super-pure silicon.

There are many other ways of producing perfect semiconductor monocrystals. But even now, from the two examples given here, it is clear that only by the concerted efforts of metallurgists and chemists, technologists and engineers it is possible to make progress in inventing and producing semiconductor devices.

SUMMARY

Some impurities (donors) are able to give away their electrons quite easily, while others (acceptors), can take electrons from semiconductor atoms. Introducing such impurities, even in very small quantities, may increase the number of free carriers to a very great extent and thus greatly increase the semiconductor conductivity.

Chapter 4. THE BIRTH, LIFE AND DEATH OF THE ELECTRON AND HOLE

> Life and death make a whole in nature.
> But it's beyond a human mind to make them one.
>
> N. Zabolotsky

We've learned much about the mysteries of birth (generation) and death (recombination) of electrons and holes in previous chapters. In this chapter we'll let you know how electrons and holes are born and how they die in semiconductor crystals which have deep impurity centers. Such impurities play a very important role in the work of many semiconductor devices. We will also describe the lives of our heroes. The life of each of them – the electron and the hole is a continuous motion.

Atoms of any substances are in constant thermal motion. Very frequent collisions of the free carriers with the atoms of the crystal lattice (they collide hundreds of billions or trillions of times per second) result in electrons and holes spending their lives in constant chaotic thermal motion. If an electric field is applied to a crystal, then the charge carriers without stopping their chaotic thermal motion also acquire a directed motion, caused by the action of the electric field. The negatively-charged electrons will drift to the positive electrode, while the positively-charged holes will drift to the negative one. The directed motion of free carriers under the action of the electric field is an electric current.

The electric current can take place, as we will soon learn, even if no external electric field is applied to a crystal. Whenever a certain part of a crystal has more free carriers (e.g. electrons) than the adjacent region, then under the action of the chaotic thermal motion the carriers begin their directed motion towards the region of a lower density. This phenomenon is called *diffusion* and the electric current created by it is called the *diffusion current.*

In this chapter we will consider the main types of motion of free carriers: the thermal motion, the motion in the electric field and diffusion.

BIRTH AND DEATH

There is a long list of impurities for any semiconductor which is commonly used, such as silicon, germanium, gallium arsenide, and indium phosphide. For silicon, for instance, this list contains several dozen names. It includes such familiar substances as arsenic and boron and also phosphorus and aluminium, silver and copper, cadmium and cobalt, gold, iron, oxygen, mercury, platinum, molybdenum, nickel, palladium, sulphur, selenium, tungsten, zinc and many other elements. Each of those elements is characterized as an impurity in silicon by its activation energy ΔE, the values of those energies lying in the range from 0.05 eV (shallow donors and acceptors) up to 0.8 – 0.9 eV. The latter values, characteristic of the "deepest" impurities, are quite comparable to the energy E_g in silicon (1.1 eV).

One might think that the "deeper" the impurity (i.e. the greater ΔE), the smaller the role it has to play. And really, it is seen from Eqs. (8) and (11) that with the increase of ΔE the density of free electrons (or holes) falls very fast, exponentially. Eqs. (8) and (11) are valid for any value ΔE. But the conclusion that the "deep" impurities can be neglected would be too hasty and quite wrong. Deep impurities in semiconductors play three very important roles: they act as "compensators", "stairs" and "killers". We will consider each of these roles in turn.

Deep Impurities: Compensators, Stairs and Killers

We will begin with the simplest role which is partly familiar to us, namely that of compensators.

Compensators. Let us imagine that we want to obtain a semiconductor, say GaAs, with a very high resistivity. Such problems are frequently facing the specialists manufacturing semiconductor devices. Films or plates whose resistivity is very high are often used as a substrate to which very thin layers of the same semiconductor with different doping impurities in it are applied.

The intrinsic semiconductor has the highest resistivity: its number of free carriers being the smallest. But what we have already learned about the impurities in semiconductors is quite enough to make us realize how difficult it is to obtain intrinsic semiconductors. Even the smallest shallow impurity density makes the density of electrons and holes increase by thousands, millions and even billions of times.

Every effort to get rid of shallow impurities entails long, complicated and expensive operations to purify the materials. But so far, appreciable results have been obtained only for germanium and silicon. For GaAs for instance, no matter how perfect the modern technology is, the shallow impurity density level is still $N \gtrsim 10^{13}$ cm^{-3}. This value is ten billion times smaller than the density of the gallium and arsenic atoms in GaAs, but it is still a million times greater than the density of intrinsic electrons and holes in the same material at room temperature (see Table 1). So, what is there to be done?

The answer is quite simple. We will not try to purify the semiconductor very thoroughly. Let it contain 10^{15} or even 10^{16} cm^{-3} of shallow donors. And we will also introduce into the semiconductor an additional impurity, whose concentration is still greater, say 10^{16} or 10^{17} cm^{-3}. But this second impurity is to be a deep acceptor.

What will happen here? As we know, electrons created by the shallow donors continue their chaotic wandering caused by thermal motion. Sooner or later, each of these electrons will approach an acceptor impurity atom which is always ready to trap the electron. The deep acceptors will take up all the electrons created by the shallow donors and it will not be easy now to release the electrons. The activation energy of these electrons from deep impurities ΔE is very great.

With the acceptor impurity being introduced, the number of free electrons decreases abruptly. Thus, introducing an acceptor impurity actually compensates the presence of the donor impurity. It would not matter at all if the amount of deep acceptors is even several times greater than that of the shallow donors. All the electrons created by shallow donors are sure to be taken up (captured). But what about the "extra" acceptors?

They are capable of taking up electrons from the semiconductor atoms and of forming holes. But they are *deep*, their activation energy ΔE is great and correspondingly, the hole concentration will be not great at all.

In the case of GaAs either chromium or oxygen is introduced into GaAs, which has not been very thoroughly purified. If before the incorporation of a deep impurity GaAs was the n-type, chromium is introduced. Chromium is a deep acceptor in GaAs. (If the initial GaAs was p-type, then oxygen is introduced; it is a deep donor in GaAs.)* This results in the obtaining of monocrystals of GaAs with a very high resistivity up to 10^7 Ohm cm. The substrates of the "semi-insulating" gallium arsenide are widely used when manufacturing GaAs devices.

So, in the first of these cases considered by us, deep impurities play the role of a positive hero. But ... "our merits and demerits go hand in hand". This wise saying refers not only to people. It is also useful when considering the properties of deep impurities. Let us imagine another situation: we want to obtain not a semi-isolating substrate, but a material of, say, n-type, as pure as possible. We keep on introducing shallow donor impurities into the crystal, and the semiconductor ... remains highly resistant. It is quite easy to guess that the reason for it is the presence of a deep acceptor in the semiconductor. And until we "overcompensate" it i.e. until the introduced shallow impurity exceeds the amount of the deep acceptors, the rise of the doping will not change anything.

Though the semiconductor is not at all indifferent to the general amount of the impurity introduced into it. If that amount is too large, then there appear various structural imperfections, the velocity of the motion of electrons and holes in the electric field decreases and other undesirable phenomena appear. Therefore, if we want to obtain pure crystals with a certain preassigned concentration of electrons (or holes), we must first of all get rid of the deep impurities wherever necessary. Technologists can cope with this

*Deep donors compensate the shallow acceptor impurity just as the deep acceptors compensate the shallow donor impurity.

task very well when dealing with the "oldest" semiconductor materials – Ge and Si. Good progress has been made of late in the technology of GaAs. But as for the presence of deep impurities in silicon carbide, the minimal level attainable is $\sim 10^{16}$ cm^{-3}.

So in their first role, that of compensators, the deep levels hinder the appearance of free carriers – electrons and holes.

Stairs. In their second role, that of *stairs* – the deep centers have just the opposite function: they facilitate the birth of electrons and holes.

We know that it takes the energy E_g for the appearance of an electron and a hole in an intrinsic semiconductor. Figure 18(a) gives a schematic diagram illustrating a process of electron-hole pair generation. Let us assume that when the electron is in the orbit, linking the silicon atoms, its energy is E_1. In order to release the electron and form a conduction electron and a hole it is necessary to overcome the energy barrier E_g.

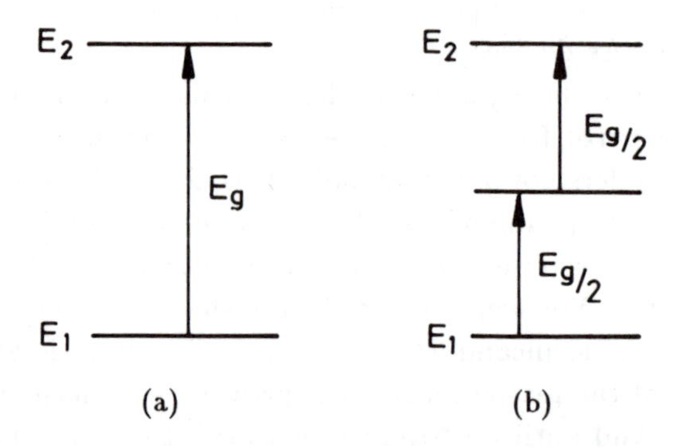

Fig. 18. Schematic diagram of the electron-hole pair generation. (a) The barrier is being overcome at one go. (b) The same barrier is being overcome in two steps.

If the energy E_g is to be obtained from the thermal fluctuation of the lattice, and the value E_g is much greater than the mean thermal energy kT, then this process is hardly probable. As we know, it is proportional to exp $(-E_g/kT)$ (see Eq. (1)). But is it possible to expend the energy necessary to release the electron not all at once but in two steps, as shown in Fig. 18(b)? First to impart the first portion of the energy, i.e. $E_g/2$, and then the second portion, also equal to $E_g/2$. It's easy to see that if we managed to do this, the process of generating electrons and holes would be much more intensive.

Perhaps the following comparison might illustrate this idea. Let us imagine that it is we and not the electron who must overcome a high barrier at one go, without any stairs. If the barrier is two meters high, then from several billion people inhabiting our planet, only a few dozen will be able to overcome it. But the same barrier divided into two parts, each one meter high, could be easily overcome by many more people. Perhaps every tenth man would be able to do it. (Fig. 19).

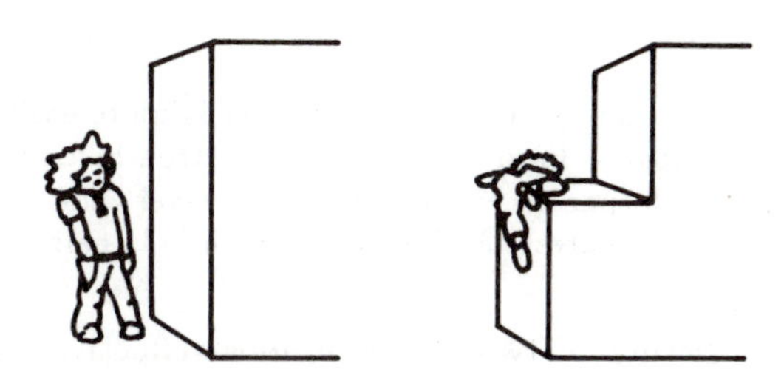

Fig. 19. It is much easier to ascend two stairs, one meter high each, than to overcome at one go a barrier two meters high.

We will now see that the deep levels whose activation energy is $E_g/2$ are in fact those "stairs", those "levels", that are shown in Fig. 18(b).

Let us assume that the acceptor impurity centers whose activation energy is $\Delta E = E_g/2$ are introduced into the semiconductor. The acceptors will take away the electrons from the neighboring atoms of the semiconductor. Thus the holes formed will begin moving chaotically in the crystal. Earlier, when we examined the behavior of the shallow acceptor, we did not think of the electron taken by the acceptor atom. And that was right. The small activation energy ΔE corresponds to a very small step up the energy stairs (Fig. 20(a)). To release the electron from the shallow acceptor and make it free requires a very large energy, practically equal to E_g. It is quite different now, when the electron is taken up by a deep acceptor. Thermal fluctuations of the lattice are most likely to break the electron from the acceptor and make it free. As a result of such a "two-step" process, an electron-hole pair has been created in the crystal. The acceptor has again acquired an ability to take away an electron from the neighboring atom of the semiconductor and form the next hole, etc. The generation rate of electron-hole pairs increases abruptly.

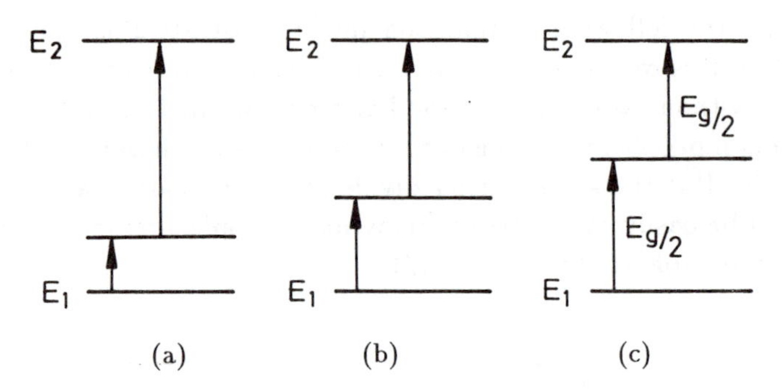

Fig. 20(a). A shallow acceptor level. The electron is quite easily trapped there. But to release it is almost as hard as to form an electron-hole pair in the intrinsic semiconductor. (b) A deeper acceptor level. (c) A level whose activation energy $\Delta E = E_g/2$ provides the greatest rate of the electron-hole pairs generation.

A process consisting of two stages is most effective for the impurity whose activation energy is $E_g/2$.

Let us assume a semiconductor material to have an acceptor level. If the acceptor's activation energy is less than $E_g/2$, then the electron from the level E_1 will jump to it quite easily. But it will be rather hard to transfer it to the level E_2 (that is to make it free) (Fig. 20(b)). Here is an exact analogy with a staircase: if you are to climb up a height of two meters and the only stair is 30 cm high, it will be quite easy to climb it. But as for the remaining 1.7m, it would be almost as hard to climb it as the original two meter height. The optimal version would be to have two stairs of equal height (Fig. 20(c)).

So introducing deep impurities greatly increases the rate of generating electron-hole pairs. But just a minute! What about the impurity properties? What about Eq. (10), from which it follows that the product of electron and hole densities depends for each semiconductor only on the temperature and not on the impurities introduced into it? What about Eqs. (8) and (11), from which it follows that the greater the activation energy ΔE, the smaller the concentration of electrons (or holes) created by that impurity, the values of N_d and T being the same?

But still there are no grounds for worry. Equations (8), (10) and (11) are valid for any level – both deep and shallow. Those equations define the electron and hole density values established as a result of two competing processes – generation and recombination.

Introducing deep centers into a semiconductor sharply increases the rate of generating electrons and holes. But the rate of recombining through those impurity centers increases as sharply too. As a result of which Eqs. (8), (10), and (11) hold true and give correct descriptions of the equilibrium densities of electrons and holes.

There are two questions with regard to this statement.

1) Why do the deep centers increase the generation and recombination rate of electrons and holes?

2) If introducing deep impurities increases both the generation and recombination rates of electrons and holes, so that their equilibrium density is determined by the activation energy alone, then what does it matter if there are any deep centers in a crystal at all?

The answer to the first question will be given a bit later on when we consider the deep centers' activities as those of "killers".

As for the answer to the second question, it comes down to the following: certain conditions are necessary for the centers to manifest themselves as stairs.

So far we have been speaking about the simplest situation, the *equilibrium* situation where the number of carriers created by thermal motion is exactly the same as that of the carriers perishing during that time due to recombination. The impurities cannot manifest themselves as stairs under those conditions. Thanks to their presence the increase of the rate of carrier generation is absolutely the same as the increase of the rate of their recombination. The carriers' equilibrium density is determined by Eqs. (8), (10) and (11).

To make the activities of the centers-stairs explicit, it is necessary to break the equilibrium between the processes of generation and recombination, creating a *nonequilibrium situation.* The best way to do it is by creating such conditions when the process of generation will go on under the same conditions of equilibrium, while the process of recombination will not take place at all. How can we manage to do it?

It proves to be not so difficult after all. It is necessary to create a strong electric field in some part of the semiconductor. Then the free carriers – both the electrons and holes – will be carried away from that region very rapidly and their density in that region will become very small. So the rate of recombination, being proportional to the product of the densities of both electrons and holes (see Eq. (3)) will diminish sharply, there being nothing to recombine. And what about the generation process? It is affected only weakly by the electric field. The electric field, able to deliver the region of a semiconductor from any free carriers, will have practically no influence on the generation rate.

The electric field removing the free carriers stipulates the appearance of an electric current in the circuit. The more free carriers generated in the semiconductor, the stronger will be the electric current. The current in a circuit is directly proportional to the generation rate of electron-hole pairs.

Let a region of a strong field be first created in a semiconductor in which there are no deep centers. Let us measure the current flowing under such conditions. Then we will introduce deep centers into the semiconductor and will measure the current again. Repeating such experiments we will make sure that the presence of even a small density ($\sim 10^{11} - 10^{12}$ cm^{-3}) of the impurities whose activation energy is $\Delta E \cong E_g/2$ increases the current tens of thousands of times. On the other hand, such impurities whose activation energy differs distinctly from the value $E_g/2$ do not have much influence on the current. Shallow donors and acceptors practically do not influence the generation rate of the electron-hole pairs.

So if the situation is distinctly nonequilibrium, when electrons and holes do not recombine, the deep centers can show their best in the role of energy stairs facilitating the birth of electron-hole pairs.

In any semiconductor device – either a diode or a transistor – at a certain regime of work there is a region of a strong electric field. The current flowing through this device at the given regime is determined by the densities of the deep levels introduced into the semiconductor.

The discussion of a nonequilibrium situation where the deep centers act as stairs can prompt the way of forming the conditions under which the same deep centers can act as killers.

Killers. Let us illuminate a specimen of the semiconductor material we are investigating with the light whose quantum energy $E_{ph} = h\nu$ is greater than the energy of creating an electron-hole pair in the material E_g. There will be excess electrons and holes in the specimen (compared to those in equilibrium) and the specimen conductance will increase.

We switch off the light. It is clear that soon the density of electrons and holes will reach the equilibrium value. The excess electrons and holes recombine, "die out". Why?

The answer is clear: after the light has been switched off the balance between the generation and recombination is broken. The generation processes will create a certain number of carriers per unit of volume and per unit of time. In the state of equilibrium the number of newborn carriers is the same as the number of carriers perishing on account of recombination. But immediately after the light was switched off, excess carriers were still to be found in the specimen and the number of both electrons and holes is still greater than in the state of equilibrium. The rate of recombination is proportional to the number of electrons and holes per unit of volume (see Eq. (3)). So the electrons and holes perish more quickly than they appear. The density of the carriers will get smaller. It will diminish until it reaches the state of equilibrium and then the rate of generation will again become equal to the rate of recombination.

The time it takes the density to return to the state of equilibrium is determined by the *lifetime* τ of the excess carriers. The sooner the density returns to its equilibrium value, the shorter is the lifetime of carriers, and the faster is the recombination process.

Measuring the lifetime τ in specimens with different impurities, we can study the influence of various impurity centers on the recombination rate.

Introducing shallow impurity centers will increase the carriers' equilibrium density in the specimen but will not influence their lifetime.

But introducing deep impurities whose activation energy ΔE approaches $E_g/2$ will immediately expose them as "killers". These impurities even though their density is very small can shorten the lifetime by hundreds and thousands of times. Thus introducing gold or platinum in a concentration of $\sim 10^{13}$ cm^{-3} (0.000 001%) into silicon accelerates the death of the nonequilibrium carriers by one thousand times: their lifetime in silicon shortens from $\sim 10^{-3}$ up to 10^{-6} s.

But even after having fully exposed the killers, a good investigator does not consider the case closed. It is necessary to answer the question "why?". Why has the crime been committed?

Let us imagine an electron and hole wandering in the crystal. In order to meet, recombine and disappear it is necessary that they should be close to each other in the vicinity of one and the same atom of the crystal lattice. Such a situation is in general possible but it seldom comes about.

Let us now assume that there is an impurity center in the crystal whose activation energy ΔE is great. Should an electron appear in the vicinity of this center, it is sure to be trapped by the impurity center. If it were a shallow center ($\Delta E \lesssim kT$), the thermal motion would almost immediately release the trapped electron. If the center were deeper (i.e. if the values ΔE were greater) then a very curious situation might exist. After having stayed for some time at the impurity center, the electron will be extracted by the thermal motion energy and will become free before the center has time to trap a hole. In this case impurity centers are called trapping centers.

And finally the center that is really deep, the *recombination* center will keep the electron trapped until a hole appears in the vicinity. As soon as that happens the electron and hole recombine. The killer has committed this task of annihilating the electron-hole pair, and ... is ready to begin all over again.

Sometimes it is important for electrons and holes to perish in the device as soon as possible. It is often quite essential for the fast switching of semiconductor devices. Then impurities creating effective recombination centers should be incorporated into the material. Sometimes on the contrary electrons and holes must live long. In this case the semiconductor is to be thoroughly purified.

We must also mention that sometimes it is impossible to get rid of the recombination centers, no matter how hard we try to purify the material. It is often not only the impurity atoms but also various structural defects of the crystal lattice of the semiconductor that can act as effective recombination centers.

Technologists are aware that if the semiconductor is heated to a high temperature and then cooled very quickly, it can diminish the lifetime of nonequilibrium carriers tens and hundreds of times even if the heating and cooling take place under conditions free from contamination. This is because when the cooling is swift there is internal stress in the crystal, since in different parts of the crystal the rate of cooling is different. As a result, there are various structural defects (thermal defects) in the lattice having all the properties of the deep centers. These defects can either take up the electrons from the neighboring atoms of the semiconductor (thermal acceptors) or facilitate the birth of free electrons (thermal donors).

Deep center properties are not only characteristic of thermal defects. For every well-known semiconductor there is a long list of lattice defects with the properties of deep recombination centers. Sometimes even a vacancy can be a recombination center. (By a vacancy we mean an empty place, the absence of an atom in the crystal lattice in the place it should be.)

On the face of it, ascribing the properties of a donor (or acceptor) to an empty place might seem somewhat mysterious and magical, but there is, of course, nothing mysterious here. The lattice atom not being in its proper place changes the conditions of the bonding between the valent electrons of the neighboring atoms, and can either facilitate the electron's breaking away from one of the atoms, nearest to the vacancy, or can incite the atom to take up an electron from the neighbors. In the first case the vacancy plays the role of a donor, in the second case that of an acceptor.

So to prolong the lifespan of electrons and holes it is necessary that the crystal should be purified and that the crystal lattice should be perfect.

The Earliest Semiconductor Devices

We have covered in this chapter a rather difficult and tiresome part. Now we deserve a short rest. We have already discussed why the conductance of the illuminated semiconductor increases, and why this excess conductance disappears as soon as the light is switched off. So, it is interesting to recall that the principle by which the earliest semiconductor devices operated was that of the conductance of a semiconductor under the influence of illumination.

In 1873, W. Smith, a London engineer, reported an unusual phenomenon discovered by his assistant A. May. While measuring the resistance of the selenium insulation of a telegraph cable, May noticed that whenever selenium was illuminated, its resistance always decreased to a great extent. The change of resistance was noticeable even when it was illuminated by dim moonlight.*

The first man to appreciate May's discovery and to make practical use of it was an outstanding German inventor Verner von Siemens, after whom a unit of conductivity was named. Siemens was the inventor of dozens of things, many of which (e.g. the dynamo) have not lost significance to this day. As far back as in 1875, Siemens manufactured the first semiconductor device – *the photometer*, made on the base of the selenium element.

The photometer's design is very simple. The greater the intensity of the light incident on the element, the greater the deviation of the galvanometer pointer, the galvanometer being connected in series with the selenium element to the circuit. Thus the photometer enables us to provide an objective measuring of the relative luminous intensity of various sources of light. So until Siemens invented the photometer, it was done just by the naked eye.

Perhaps we could appreciate this invention better if we recall two circumstances.

---

*So, selenium quite unexpectedly justified the celestial name given to it as far back as 1817 by the Swedish chemist J. Berzelius. Having discovered in 1817 a new element, Berzelius named it after the Moon (the Greek word for moon is *Selena*).

1) Modern photographic exposure meters manufactured today all over the world are in fact nothing else but Siemen's photometers. And now they are produced in the millions.
2) The comparison of the brightness of various stars and other celestial bodies is a very important problem of astronomy, one of the most ancient sciences. The Siemens photometer enables objective and exact photometry instead of comparing the brightness of the stellar light by sight. So thanks to that invention, a very important step has been made in developing astronomy and making it an exact science based on quantitative measurement.

In 1878 , three years after the invention of the photometer, Alexander Graham Bell, a famous American inventor and creator of the telephone, informed the world of his having combined in one device the two most wonderful inventions of the 19th century – the telephone and the photometer. Bell constructed a new device, the photophone, intended to realize the idea of a wireless telephone.

The first wireless telephone communication in history took place between Bell and his assistant Teinter in the summer of 1878. Teinter was on the roof of the Franklin school in Washington where the "transmitting telephone station" was placed. It was designed in the following way. A ray of the sun was focused by means of lenses on a small mirror, connected on the rear side to a megaphone. The ray, reflected from the mirror, was directed to the window of Bell's research laboratory which was opposite the school at a distance of 213 m. There was a "receiving station" there containing a selenium photoelement, connected to a battery and to a telephone. Approaching the megaphone, Teinter pronounced "Mr. Bell, if you hear me, come up to the window and wave your hat". The megaphone's vibration was transferred to the mirror, the intensity of the reflected ray began changing "in time" with Teinter's words, and the selenium element transformed light oscillations into the electric oscillations. Bell came up to the open window and happily waved his hat. The sensation caused by the invention of the photophone was great but short. The ordinary telephone proved to be simpler, cheaper and more reliable. The photophone had no practical use at all. For a long time, but not forever.

Recently the idea of a photophone has been revived but at a higher technical level. It has been incorporated into the systems of fiber optical communication. Just as in Bell's photophone the signal from the transmitting station is now sent to the receiving station by means of a ray of light. But the light is not sent through the air where it would shortly be attenuated. It is sent through the optical fiber where it can propagate quite easily without being weakened and can thus travel a very long distance. The source of light is also different. It is not sunlight but semiconductor lasers which are used as the source of light. The luminous intensity is changed by semiconductor modulators with a frequency of billions of times per second. At both the transmitting and the receiving stations the electric signals are treated instantly by special computers. Nowadays there are networks of cables of fiber optical communication in almost all industrially developed countries. They are tens of thousands of kilometers long and transmit annually hundreds of millions of telephone messages.

LIFE AND MOTION

Thermal Motion

There is always a chaotic thermal motion in nature. The higher the temperature of the crystal, the greater the energy of thermal motion. The average energy of the thermal motion of electrons (or holes) equals 3/2 kT. Equating this value to the kinetic energy of the particle $mv^2/2$ we can find the average velocity v_T of the chaotic motion of the electron (or the hole):

$$v_T = \left(\frac{3kT}{m}\right)^{1/2} . \tag{12}$$

When $T = 300$ K, assuming that the mass of the free carrier is equal to the mass of the free electron, we find that the velocity v_T is $\sim 10^5$ m/s.

One hundred kilometers per second! It is ten times greater than the Solar escape velocity. A body whose velocity is as great as that will leave the Solar system. But for an electron such a velocity is insignificant – the velocity which the electron receives in the accelerators is 3000 times greater and is practically equal to the velocity of light. As for the velocity $v_T \sim 10^5$ m/s, it corresponds to the electron kinetic energy equal to several hundredths of an electron-volt. It is not sufficient for the electron even to leave the semiconductor crystal, to flow out beyond it. (To perform such an "exploit" the electron would require the energy equal to several tenths of an electron-volt or even to several electron-volts).

Motion in the Electric Field

A drift of particles directed along the lines of the electric field is added to the chaotic thermal motion of free carriers (electrons and holes). Let us determine the velocity of the directed motion.

If there is no electric field, then the electron (or the hole) takes part only in the chaotic thermal motion. There is no directed motion whatsoever, there being no such direction which the electron (the hole) would prefer. So, although the electron (the hole) moves chaotically with a very great velocity $\sim 10^5$ m/s, as we have established it in the previous item, the velocity of the directed motion equals zero.

In the presence of the electric field the electrons (the holes) continue their mad dance. The frequency of their collisions is, as before, $\sim 10^{12}$ – 10^{13} times per second. It is still impossible to foretell where the carrier will flow after the next collision, whether forward or backward, to the right or to the left, upward or downward, at what angle and in what direction. But no matter where it flows, the electric field will always draw it, though perhaps quite weakly, in the same direction, which will result in the appearance of a directed motion.

In the electric field $\mathbf{F}$ the electron (or the hole) is acted upon by a force $f = qF$. Under the action of this force the carrier acquires an acceleration $\mathbf{a} = q\mathbf{F}/m$ along the line of the field. Moving without any collisions, during the time t the carrier will acquire a velocity in the direction of the field $\mathbf{v} = \mathbf{a}t = q\mathbf{F}t/m$.

In order to calculate the average velocity acquired by the carrier under the conditions of repeated collisions we must remember two circumstances.

In the first place, as we know, after a collision the carrier can move in any direction. That means that the velocity of the *directed* motion after the collison is equal to zero. In the second place, since the collisions are quite accidental, the time of the carrier's "free flight" can also be quite different. This can be clearly seen in Fig. 21, showing the velocity-time dependence of the directed motion.

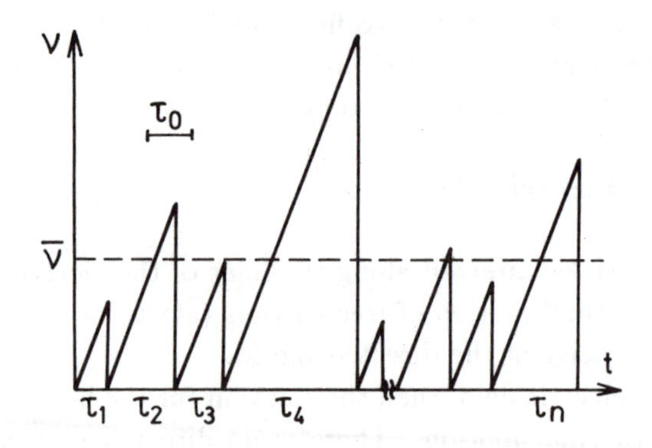

Fig. 21. The velocity of the directed motion of the electron (or the hole) in the electric field versus time at random scatterings.

The average velocity of the directed motion $\bar{v}$ is equal to the product of acceleration and the average time between the collisions τ_0

$$\bar{v} = \frac{q\tau_0}{m} F \; . \tag{13}$$

So the velocity of the directed motion which is often called the ***drift velocity*** of free carriers in a crystal is proportional to the electric field. The coefficient of the proportionality

$$\mu = \frac{q\tau_0}{m} \tag{14}$$

is called *mobility*.

Mobility. Mobility of the free carriers is one of the most important characteristics of a semiconductor crystal.

From Eqs. (13) and (14) it follows that

$$\bar{v} = \mu F \; . \tag{15}$$

The greater the mobility the greater the velocity of the directed motion of the carriers for a given electric field. From Eq. (15) it is clearly seen that the dimension of mobility is $[\mu] = \mathrm{m^2/(V \cdot s)}$.

Table 2 gives the values of the mobility of electrons and holes of the most important semiconductor materials at room temperature.

Table 2

Semiconductors	InSb	Ge	Si	InP	GaAs	GaP	SiC
Mobility of electrons μ_n $\mathrm{m^2/(Vs)}$, 300 K	8	0.39	0.13	0.55	1	0.05	0.02 – 0.1
Mobility of holes μ_p $\mathrm{m^2/(Vs)}$	0.07	0.19	0.05	0.07	0.04	0.01	$5\ 10^{-4}$

In weak fields, when the velocity of the directed motion is very small in comparison with that of the thermal motion, neither the presence nor the absence of the field can affect the character of the carriers' collisions with the crystal lattice. Mobility μ is a constant value, it does not depend on the electric field F. The drift velocity of the carriers is proportional to the field F (Eq. (15)) in accordance with Ohm's law.

Now we've made clear the main difference between the electron which is actually free and the so-called "free" electron in a semiconductor crystal, the conduction electron. In the crystal, the chaotic collisions being rather frequent, it is not the acceleration but the velocity of the conduction electron (or the hole) that is proportional to the field.

Let us make sure that the situation where the velocity of the electron is proportional to the field is in accordance with Ohm's law. Let us recall some relations which we must know from a school course in physics.

Ohm's law states: $I = U/R$.

The resistance R depends on the length of the resistor L, on its cross section S and resistivity ρ: $R = \rho L/S$. Although in semiconductor physics the inverse value of resistivity – conductivity – is used preferably. Hence, to express the resistance R the following expression is used

$$R = L/(\sigma S) .$$

The intensity of the field in the sample is $F = U/L$. The current I flowing through the resistor is equal to $I = j \cdot S$, where j is the current density related to the density n_o and the drift velocity $\overline{v}$ by the well-known ratio $j = qn_o\overline{v}$. Hence, $I = qn_o\overline{v}S$.

Substituting into the equation of Ohm's law the expressions for the field, resistance and current, we obtain

$$\overline{v} = \frac{\sigma}{qn_o} F .$$

So the proportionality of the drift velocity of the carriers to the electric field follows directly from Ohm's law. Comparing the expression we obtained with Eq. (15), we see that $\mu = \sigma/qn_o$ i.e. the conductivity of the material σ is directly proportional to the mobility of the carriers in it.

From Eq. (15) and the expression for the current density we can draw another very useful relation

$$j = \sigma F . \tag{16}$$

The current density at any point is proportional to the field F. Equation (16) is often called Ohm's law in its differential form.

At first sight it may seem we have made a thorough study of what mobility is and that now we may go on. But let's not hurry. There are a few questions which inevitably arise.

First of all, what is meant by τ_0 in Eqs. (13) and (14)? Is that parameter defined by the time between two collisions of the carrier? If it is so then it is quite natural to assume it to be the time the carrier travels between the two neighboring atoms of the lattice. Then we can easily calculate the value τ_0 and consequently the mobility value μ.

Let us recall that while moving under the action of the electric field, the electron (or the hole) does not stop its chaotic thermal motion. The velocity of the thermal motion v_T is, as we have seen, equal to 10^5 m/s. The velocity of the directed motion under the action of the field is as a rule much smaller. Therefore the time it takes the carrier to fly between the two neighboring atoms is determined by the velocity v_T.

The distance between the two neighboring atoms of the lattice a_0 is approximately the same in every solid and is known to be equal to $\sim 5 \cdot 10^{-10}$ m. Moving with thermal velocity $v_T \sim 10^5$ m/s, the electron covers the distance between the atoms in an average time of $\tau_0 \sim 5 \cdot 10^{-15}$ s. According to Eq. (14) this value τ_0 answers the mobility value $\mu \sim 10^{-3}$ m^2/(Vs).

Since the values a_0 and v_T are approximately the same for any crystal, the mobility μ, according to our idea of the nature of the collisions, must also be approximately the same for any solid body.

Let us look again at Table 2 and compare our theoretical predictions with the experimental results.

Oh, dear! The comparison is quite discouraging. The mobility values of various semiconductors differ by thousands and tens of thousands of times, and in some cases by a factor of 10^4 or 10^5 greater than the value we have forecasted. It is quite obvious now that as theorists we proved ourselves to be inadequate.

Well, let us try to approach the problem from another point of view. Starting with the mobility values obtained experimentally, let us calculate the distance the electron must cover before a collision. Perhaps we will have a better idea then of what it collides with.

Let us take InSb as an example. From Eq. (14) the value $\tau_0 \sim 5 \cdot 10^{-11}$ s corresponds to the magnitude $\mu = 8$ m^2/Vs. Provided the thermal velocity is $v_T \sim 10^5$ m/s, between two collisions the electron covers the way $l \sim v_T \cdot \tau_0 \sim 5 \cdot 10^{-6}$ m. Since the distance between the atoms is of the order of $5 \cdot 10^{-10}$ m it corresponds to 10 000 interatomic distances!

This seems really unbelievable. Before colliding (with what?), the electron passes 10 000 atoms. The old wisdom has it that it is easier for a camel to pass through a needle's eye than for a rich man to get to heaven.

But it sounds easier for a rich man to get to heaven than for an electron to fly past ten thousand atoms and not to collide with any of them!

Nevertheless that's how things are.

How can such miracles be interpreted? No interpretation can be given if the electron and the hole are assumed to be particles, balls, which are to break through a paling of tens of thousands of atoms without colliding with any of them. But what seems impossible for a particle appears quite possible and even natural if we take into consideration the wave nature of the electron.

In accordance with the main ideas of quantum mechanics, the electron (or the hole), like any other particle has not only corpuscular but also wave properties. With regard to light this dualism does not seem strange. We are accustomed to it. While analyzing the phenomena of diffraction or interference, we emphasize the wave nature of light. Discussing the phenomenon of photoeffect, we assume the beam of light to be a stream of particles, of photons. We are not accustomed to assume that the electrons are waves. But the fact that the electron possesses the properties not only of a particle but also of a wave has been proved most convincingly, as convincingly in fact as for the light. Right at the dawning of developments in quantum mechanics Devisson's and Germer's wonderful experiments proved the phenomenon of electron diffraction.

Under certain conditions any kind of wave (accoustic, light, or radio waves) can propagate without being reflected, in a medium containing some scattering centers. The main condition of this propagation is a regular arrangement of the scattering centers, that is the centers should be arranged as an ideal periodic lattice. The slightest deviation from the ideal spacing: the change of the distance between the centers, the substitution of one center for another whose properties are but slightly different, even an absence of a center in the place where it should be in the lattice causes the scattering of the wave.

And the electron-wave too, just like any other wave, while in an ideal crystal which does not contain any impurity atoms or any vacancies (empty places in the lattice), or any distortion when the temperature is at absolutely zero, would not experience any scattering collisions and would travel quite freely.

Scattering Mechanisms. And yet what does the electron collide with? The time denoted by τ_0 in Eq. (14) is between which collisions? We seem now to have found the correct answer. τ_0 denotes the time between the electron's collisions with any distortion in the ideal periodic lattice of the crystal. If the crystal contains any impurities, no matter how negligible their density might be, the electron wave will be scattered by an impact with them and the collision time τ_0 will be determined by the impurity density (the so-called *impurity scattering*). The same will be true if the crystal contains any structural imperfections, any empty sites of the crystal lattice and so on. If the temperature of the crystal is not zero, the atoms forming the crystal lattice vibrate chaotically and at any moment some atoms will be closer to each other than in the ideal lattice when the temperature was $T = 0°$K and some atoms will be further from each other. The ideal spacing will be broken. There will be scattering when atoms of the lattice vibrate (the so-called scattering by *lattice vibrations*). The higher the temperature, i.e. the greater the vibration amplitude, the larger the scattering.

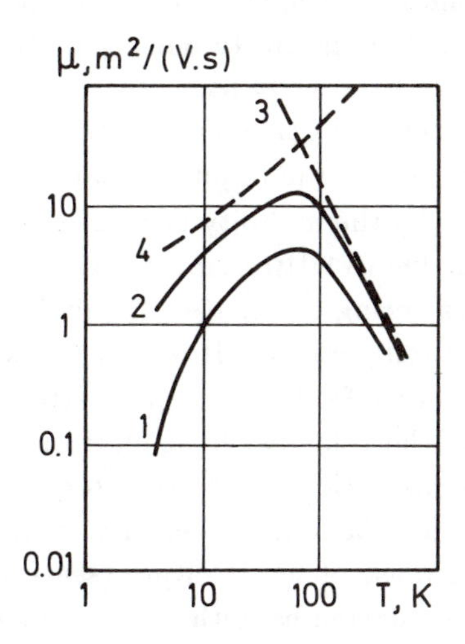

Fig. 22. Mobility versus T in gallium arsenide containing various impurity atom densities.
1 – the shallow donor density $N_d = 2.5\ 10^{14}\ \text{cm}^{-3}$;
2 – $N_d = 6.5\ 10^{13}\ \text{cm}^{-3}$. The broken lines indicate theoretic dependence with scattering by lattice oscillations only (curve 3) or by impurity centers only (curve 4).

One can learn a lot about the causes (they are often called the *mechanism*) of scattering by studying the mobility-temperature dependence. Figure 22 shows the mobility – temperature dependence of gallium arsenide containing different impurity atom densities. This dependence is typical of many other semiconductors as well.

It is seen from the curves that when the temperature is high, mobility in practice does not depend on the impurity density at all. It is determined by the scattering of the electrons on the thermal lattice vibrations. The higher the temperature, the stronger the atomic vibrations, and the lower the electron mobility.

If there were no impurities or imperfections in the crystal, then with the lowering of the temperature the mobility of the carriers would increase infinitely. The mobility-temperature dependence $\mu(T)$ would be determined by curve 3 in Fig. 22. But it is absolutely impossible to obtain a crystal which would be without any impurities or defects. Figure 22 shows that in reality the mobility first increases and then on reaching its maximum begins to fall.

It is seen that in the low temperature region mobility is very sensitive to the impurity density. The impurity density of 10^{17} impurity atoms per 1 cm^3 of GaAs (i.e. only 1 impurity atom to 100 000 host atoms) reduced the mobility by a factor of 5, the temperature being 10 K. It is obvious that in the low temperature region the impurity scattering is dominant.

We do understand why the mobility decreases with the rise of temperature when scattering is due to lattice vibrations. But it is seen from Fig. 22 that when impurity scattering dominates, the mobility on the contrary increases with the rise of temperature. How can we explain it?

Figure 23 gives a scheme for the electron scattering by a charged impurity atom. The greater the velocity of the electron flying past an impurity atom, the smaller the distortion of its trajectory under the action of the electric field of the atom. The average velocity of the electron thermal motion grows with the increase of temperature (see Eq. (12)). Therefore the scattering by impurities decreases with the rise of temperature and as a result the mobility increases.

Let us return to Eq. (14). There are only three quantities in the right-hand side determining the mobility. Examining one of them τ_0, we can see how deceptive is the simplicity of that equation.

Effective Mass. At the beginning of this section when we naively assumed the electron to be flying between the two collisions in the empty interatomic space, we had every reason to believe the mass m to be the free electron mass. But now we know that things are not so simple as they seem

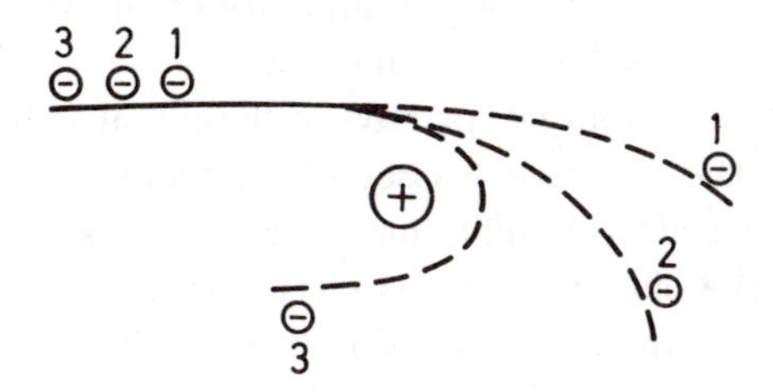

Fig. 23. The carrier scattering by a charged impurity center. Electrons (1, 2 and 3) fly up to the impurity center along the same trajectory but with different velocities. Electron 1 has the greatest velocity; the attraction of the impurity atom in practice does not affect its initial trajectory. Electron 3 has the lowest velocity; the attraction of the charged impurity makes it change its route backwards.

to be. In the time interval between two acts of scattering, the electron and the hole can pass distances measured by thousands and tens of thousands of the lattice constants. That means that while moving freely the electron and the hole are acted upon by the external electric field F and also by very strong fields created by the crystal lattice ions and by the valence electrons of the atoms between which the free carriers are flying. That cannot fail to affect the motion of electrons and holes. It is rightful to ask: does the reasoning that brought us to derive Eqs. (13) and (14) make sense? And, consequently, can we use those equations?

The answer to these questions can be obtained only from quantum mechanics which considers both the wave and the corpuscular properties of electrons and holes.* The answer is the following:

The electric fields of the crystal lattice atoms being arranged in a strictly periodic order, Eq. (13) can be used to define the motion of electrons and holes in the semiconductor. However, the mass m in the denominators of Eqs. (13) and (14) ought to be interpreted not as the free electron mass in vacuum m_0, but as a value depending on the type of the semiconductor and called "the effective mass" of the electron or the hole. The symbol m^* is usually used to denote it. The replacement of the free electron mass m_0

---

*A long time ago the great German poet Heinrich Heine wrote that the difference in Latin between the regular verbs and the irregular verbs lies in the fact that students are more often flogged for the latter than for the former. There are students who are sure that the difference between quantum and classical mechanics is the same. This is correct to a certain extent. But there are some other differences between them.

by the effective mass m^* reflects the influence of the strictly periodic field of the crystal lattice on the electron motion.

Since different semiconductors have different atomic parameters and lattice structures, the effective masses of electrons and holes in different semiconductors must also be different. And so they are.

The values of the effective masses of electrons m_e^* and of holes m_h^* in some semiconductors are given in Table 3. Keep in mind that the figures given there represent the ratio of the effective mass of an electron (or a hole) to the free electron mass in vacuum $m_0 = 9.1\ 10^{-31}$ kg.

The values m_e^* and m_h^* are seen to be quite different from the free electron mass m_0. The electron effective mass in indium antimonide (InSb), for instance is almost 80 times smaller than the free electron mass! No wonder the electron mobility values for InSb are so high (see Table 2).

Table 3

Semiconductors	InSb	Ge	Si	InP	GaAs	GaP	SiC
m_e^*/m_0	0.013	0.12	0.26	0.07	0.07	0.35	0.6
m_h^*/m_0	0.18	0.28	0.49	0.2	0.45	0.86	1.2

Since the effective masses of electrons and holes vary greatly in different semiconductors, it is clear that the thermal velocities of the free carriers, defined by Eq. (12) are also different, provided temperature T is the same. For $T = 300$ K the value of the average electron thermal velocity is $\sim 10^6$ m/s in InSb, $\sim 4.5\ 10^5$ m/s in GaAs and $\sim 1.5\ 10^5$ m/s in SiC. At room temperature, the hole thermal velocities in the same materials will be $2.5\ 10^5$, $1.7\ 10^5$ and 10^5 m/s accordingly.

Hot Carriers. Now, when we have specified the velocity values of the disordered chaotic motion of carriers in various materials and have studied rather thoroughly the notion of mobility, let us calculate what electric field is needed to make the velocity of the directed motion of the carriers equal to the thermal velocity v_T. The corresponding value of the field F_0, which is apparently equal to $F_0 = v_T/\mu$ will at room temperature make $\sim 10^5$ V/m for n-InSb, $\sim 5\ 10^5$ V/m for n-GaAs and $\sim 10^6$ V/m for n-SiC. For p-type semiconductors the corresponding values of F_0 are still higher since the mobility of holes is as a rule lower than that of electrons. (See Table 2).

To better understand the electric field value of $\sim 5 \cdot 10^5$ V/m we should note that such a field will be created in a sample which is 1 cm long if a voltage of 5 000 V is applied to it.

If the field applied to a semiconductor is so great that the velocity of the directed motion of the carriers approaches the velocity of thermal motion then the average energy of the carriers (electrons and holes) increases considerably. The carriers are said to become "hot". Their interaction with the lattice is different from that of the carriers in a weak electric field. The collision time τ_0, the effective mass m^*, and consequently, the mobility of hot electrons depend on the electric field.

Practically in all semiconductors at room temperature the mobility in strong fields decreases with the increase of the field F. In very strong fields the mobility value is inversely proportional to the field F: $\mu \sim 1/F$. In accordance with Eq. (15) that means that in very strong fields the drift velocity of the carriers does not depend on the field: $\bar{v}$ = const.

Essentially, a charge carrier (an electron or a hole) which has acquired a certain, sufficiently large energy E_0 between two collisions, completely transfers it to the lattice at every collision. Let us make sure that under such conditions the time between the collisions τ_0 will be inversely proportional to the intensity of the field F.

The energy of the carrier is defined by its velocity and mass: $E = m^* v^2/2$.

The velocity is determined by acceleration $a = qF/m^*$ and the time between the collisions τ_0: $v = qF\tau_0/m$.

Substituting the expression for v into the equation connecting the energy and velocity of the carriers we obtain

$$\tau_0 = \frac{\sqrt{2m^* E_0}}{q^2} \frac{1}{F} .$$

In accordance with Eq. (14) the value μ in such situations will also be inversely proportional to the field F. And the drift velocity of electrons $v = \mu F$ will be a constant value which does not depend on the field.

The dependences of the drift velocity on the field for Ge and Si are shown in Fig. 24. It is seen that in the weak electric fields the velocity is always proportional to the field. This part of the curve is quite reasonably called the Ohm or linear region. The stronger the field, the slower the growth of the drift velocity and when the field is strong enough the velocity ceases to depend on the field F altogether. (This is the region of a *saturated velocity*). The smaller the mobility of the carriers in a weak electric field, the stronger

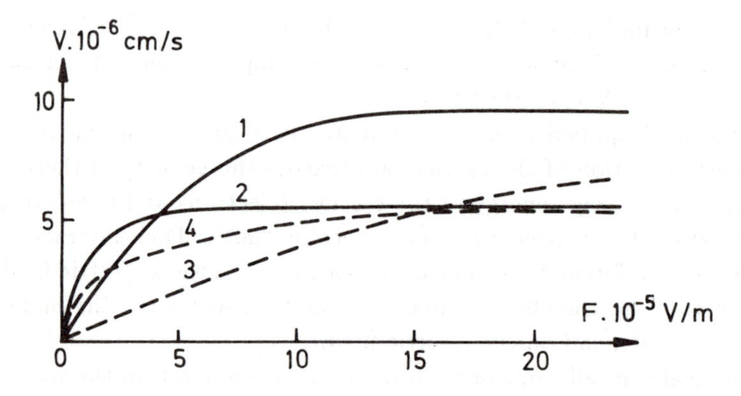

Fig. 24. The velocity of the directed motion (the drift velocity) versus the electric field in Ge and Si ($T = 300$ K). Curves 1, 3 – Si; 2, 4 – Ge. Solid lines – the electron v(F) dependences, broken lines – the hole v(F) dependences.

must be the electric field for the velocity to be saturated.* In germanium when the field is $F \gtrsim 5$ kV/cm the velocity of electrons in practice does not depend on the field any longer. The velocity of holes saturates for $F \gtrsim 15$ kV/cm. The values of the saturated velocity v_S for both the electrons and holes are approximately the same, and are equal to $\sim 5 \cdot 10^6$ cm/s. In Si the electron saturated velocity is $v_s \sim 10^7$ cm/s, it is reached for $F \gtrsim 15$ kV/cm. For holes in Si the value of the saturated velocity is also $\sim 10^7$ cm/s, but it is reached in much stronger electric fields: $F \gtrsim 60$ kV/cm. Since both values τ_0 and the effective mass of the carriers m^* depend on the carrier energy, the $\bar{v}(F)$ dependence in semiconductors can have a much more intricate look than that given in Fig. 24. Thus, in GaAs when F increases and reaches certain values, the velocity of the electrons does not increase, but falls with the increase of F. In Chapter 9, when studying the Gunn effect, we will have a better understanding of such an unusual dependence $\bar{v}(F)$.

Diffusion

The word diffusion comes from the Latin word "diffusio" (overflow, penetration). This notion can be applied to gases, liquids and solids.

The smell of the perfume spilled in a room spreads throughout the flat. Even if the radiator is switched off, the window is closed and all the chinks

*Do you understand why?

are caulked up to make the air absolutely motionless, the molecules of the perfume will eventually penetrate into other rooms due to the process of diffusion.

A thin coating of substance containing a lot of impurity is applied to the surface of a semiconductor plate thoroughly cleaned of any impurity. Some time later the applied impurity can be found in the semiconductor plate quite far from the surface. The impurity gets deep into the plate due to the process of diffusion. The rate of the impurity diffusion in a semiconductor increases very much with the increase of temperature. At room temperature it would take decades for the impurity to penetrate into the volume of a semiconductor, but when the temperature is high enough the process of diffusion can take a few hours or even a few minutes. (Incidentally, the diffusive introduction of impurities is widely used when manufacturing semiconductor devices.)

What is there in common between the phenomena described above? It is what makes the very essence of the process of diffusion – the spontaneous penetration of the substance, unaffected by any outer influence from the place where it is in a great quantity to the place where its quantity is scarce.

The process of diffusion is a direct consequence of the chaotic thermal motion of atoms (or molecules). It is very important to understand why the disordered chaotic motion of particles leads to the directed displacement of the substance from an area of high density to an area of low density. The answer is quite simple.

Let us discuss the case with the spilled perfume. Let us imagine a hemisphere surrounding the spilled liquid by which the molecules can penetrate quite easily. We will count the perfume molecules which pass through the surface of the hemisphere from within and from outside. The perfume molecules collide with the molecules of the air and scatter in different directions. Those of them which approach the surface of our imaginary hemisphere, being kicked in a certain direction, can cross the surface. If the density of the perfume molecules were equal on both sides of the hemisphere then the stream of molecules directed outside would be equal to the stream of molecules directed inside.

But in our case things are different. The density of the perfume molecules n is great at the place where the perfume was spilled and it becomes smaller the farther we move from it (Fig. 25). Outside the surface we have mentally drawn, the perfume density will be lower than inside it. Under the influence of the chaotic motion, the same fraction of the molecules which are close to the surface of our imagined hemisphere will cross it in both directions from outside as from inside. The same fraction but of a different quantity, because inside the hemisphere the density is higher than outside. Therefore a greater number of the perfume molecules will cross the surface from the inside than from the outside. A stream of molecules is formed, directed outside, the stream which appeared as a result of a chaotic motion. That is how the process of diffusion is formed, the process which results in equating the densities.

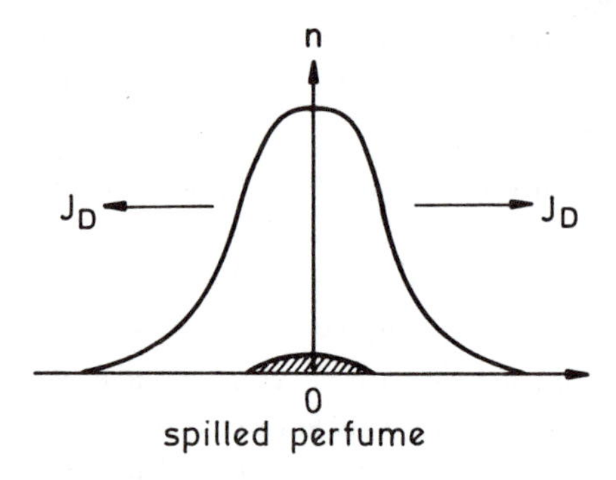

Fig. 25. The distribution of the perfume molecules over the spilt perfume. In the region $x > 0$ the derivative is $dn/dx < 0$ (n decreasing with the growth of x), the diffusive flow of carriers is $\mathcal{J}_\mathcal{D} > 0$, (along the positive direction of the x axis); in the region $x < 0$ the derivative is $dn/dx > 0$, the diffusive flow being $\mathcal{J}_\mathcal{D} < 0$.

From what has been said it is clear that the greater the density difference on both sides of our imaginary surface, the greater will be the molecular stream caused by diffusion. The density difference determines the relation between the two streams moving in opposite directions. The rate of change of density n with the x coordinate is characterized by the derivative of the density with respect to the coordinate dn/dx. Thus the diffusive flux of carriers is expected to be proportional to dn/dx

$$\mathcal{J}_\mathcal{D} = -\mathcal{D}\,\frac{dn}{dx}\,. \tag{17}$$

Why is there a minus before the right part of that equation? Figure 25 helps us to answer that question. Wherever the derivative dn/dx is positive, the flux is directed against the positive direction of the x axis and vice versa.

The proportionality factor $\mathcal{D}$ is called the *diffusion coefficient.*

Diffusion Coefficient. Let us first of all establish the dimension of the diffusion coefficient. The flux of carriers is numerically equal to the number of carriers crossing a unit of area per unit of time. Thus, the dimension of the flux is $[\mathcal{J}_\mathcal{D}] = (m^2 \cdot s)^{-1}$. The density dimension is $[n] = m^{-3}$ and the dimension of the concentration gradient $[dn/dx]$ is, evidently, m^{-4}. Hence it is clear that the dimension of the diffusion coefficient $\mathcal{D}$ is $[\mathcal{D}] = m^2/s$.

What does the diffusion coefficient depend on? By what molecular properties is it defined?

Let us derive the expression for the diffusion coefficient $\mathcal{D}$ using the so-called dimensional method. This method may seem at first sight a juggling trick or even a smart swindle, but in fact it is a well-known method, widely used in modern physics.

So what can the diffusion coefficient depend on?

It is natural to expect it to depend on the length of the free path of the molecules l.

The greater is l, the more seldom the molecule changes its path after its collisions with other molecules. In our example with the spilled perfume it is clear that if the collisions with the air molecules did not impart to the motion of the perfume molecules any character of chaotic wandering then starting from the surface of the perfume with a velocity v_T, the perfume molecule might reach the end of the room whose length is L in time $t \sim L/v_T$. But in reality it will get there much later. It will take it more time to get there. The more frequent are the collisions with air molecules, the shorter is the free path l, the more slowly the perfume molecules diffuse and the smaller is the diffusion coefficient $\mathcal{D}$.

With the growth of the thermal velocity of molecules v_T it is natural to expect the diffusion coefficient to increase.

What else can the diffusion coefficient depend on? It depends on the collision time τ_0 of course! That's right. But the value τ_0 is expressed in l and v_T: $\tau_0 = l/v_T$. Does it depend on anything else? What about the mass of the molecules? It's hardly likely. We are interested neither in the energy transferred by molecules nor in the momentum. We are interested just in the number of molecules.

So let us assume that knowing the physical picture of diffusion, we have given a correct definition of the variables on which the diffusion coefficient $\mathcal{D}$ depends. Then there is only one problem left. It is the main problem in the dimensional method, but it is very simple in our case: to combine the variables (in our case l and v_T) in such a way that the combination might have the dimension of the wanted value. The dimension of the length is $[l] = m$ and the dimensions of the velocity is $[v_T] = m/s$. It is easy to see what is to be done in order to obtain the value $\mathcal{D}$ whose dimensionality is $[m^2/s]$. We must multiply them.

$$\mathcal{D} \sim l \cdot v_T \ . \tag{18}$$

Explicit calculation enables us to get the numerical coefficient

$$\mathcal{D} = \frac{1}{3} l \cdot v_T \ . \tag{19}$$

Now let us leave alone the perfume molecules and let us return to our old friends – the free carriers in a semiconductor.

Diffusion Current. Let us assume one part of a semiconductor to have a greater carrier density than the neighboring sections. This may happen if a certain section of a semiconductor is heated or illuminated. Then as we know, due to diffusion, electrons will flow from a higher to a lower density region. But the directed flow of electrons is by definition an electric current! Knowing the distribution of the carriers $n(x)$, we can easily find the density of this diffusion current.

$$j_{\mathcal{D}} = q\mathcal{J}_{\mathcal{D}} = -q\mathcal{D}\,\frac{dn}{dx} \ . \tag{20}$$

The density of the diffusion current is equal to the flux density of the charge carriers defined by Eq. (17) multiplied by the electron charge.

When calculating the diffusion current we should bear in mind the charge of the carriers whose behavior is being studied. If the charge is negative (the electrons), then though the flux of electrons will be directed towards the region of the lower carrier density, the electron diffusion current will flow in the opposite direction, towards the higher density: $j_{\mathcal{D}n} = |q| \cdot \mathcal{D}_n \frac{dn}{dx}$. In this equation $\mathcal{D}_n$ is the diffusion coefficient of the electrons.

If the charge is positive (the holes), the direction of the diffusion current is the same as that of the hole flux towards the lower density: $j_{\mathcal{D}p} = -|q| \cdot \mathcal{D}_p \frac{dp}{dx}$, where $\mathcal{D}_p$ is the diffusion coefficient of the holes.

The diffusion current is a real current no different from the current which appears under the action of the electric field F. When passing, it produces Joule heat, just like the conduction current (i.e. the current which appears under the action of the field). The diffusion current causes the deflection of the magnetic needle. To calculate the diffusion current it is necessary to know the diffusion coefficients of both, – electrons and holes $\mathcal{D}_n$ and $\mathcal{D}_p$.

The Einstein Relation. There is a simple relation between the diffusion coefficient of any particles and their mobility. We will rewrite Eq. (19) as

$$\mathcal{D} = \frac{1}{3} v_T^2 \cdot \tau_0 \tag{21}$$

and will substitute into it the square of the thermal velocity from Eq. (12)

$$\mathcal{D} = \frac{1}{3} v_T^2 \cdot \tau_0 = \frac{kT}{m} \tau_0 \; . \tag{22}$$

Comparing Eqs. (22) and (14) for the mobility of the particles, we obtain

$$\mathcal{D} = \frac{kT}{q} \mu \; . \tag{23}$$

Knowing the value of the mobility at a given temperature, we can easily find from Eq. (23) the value of the diffusion coefficient.

The relation between the diffusion coefficient and mobility, established by Eq. (23), is valid not only for electrons and holes but also for any particles both charged and uncharged,* moving either in the gravity field or in the electric field. This relation is a vivid reflection of the fact that both the directed motion of particles under the action of force and the process of diffusion are hindered by one and the same process – by the collisions of particles which occur within mean time intervals τ_0 with the average thermal velocity of particles v_T.

---

*For the uncharged particles mobility μ is defined as the relation of the average velocity of the particles $\bar{v}$ to force f, under the action of which this velocity is established: $\mu = \bar{v}/f$. In this case the relation between the diffusion coefficient and mobility is as follows: $\mathcal{D} = kT\mu$.

The connection between the diffusion coefficient and mobility was established by Albert Einstein, and Eq. (23) is called the Einstein relation.

Substituting into Eq. (23) the Boltzmann constant k and the electron charge q, it is not difficult to make sure that at room temperature (300 K) the value kT/q is equal to about 0.026 V. Thus (see Table 2) at room temperature, the diffusion coefficient of the electrons in InSb is about 0.2 m^2/s, of the electrons in Ge – 0.01 m^2/s and in GaAs – 0.025 m^2/s.

Diffusion Velocity. Let us discuss now an important question, the velocity of diffusion. Or in other words, let us determine the time t required for the diffusive particles to cover the distance L.

To solve this problem we will use the dimensional method.

Since the answer must comprise only two values: L (in m) and $\mathcal{D}$ (in m^2/s), then we can easily see that the only way to obtain a value having the dimension of time is to divide the square of the length L by the diffusion coefficient $\mathcal{D}$. So the dimensional method gives the following answer to this question:

$$t \sim \frac{L}{\mathcal{D}} \quad \text{or} \quad L \sim \sqrt{\mathcal{D}t} \; . \tag{24}$$

Equation (24) established a most unusual distance-time dependence. Distance L is proportional not to time t but to the square root of time t. Therefore at first sight it may seem that diffusion is a slow process.

But let us estimate it numerically. The sizes of modern semiconductor devices are often of the order of a micrometer (10^{-6} m) or even a fraction of a micrometer. How long will it take an electron moving on account of diffusion to cover a distance of 10^{-6} m? As we know, at 300 K the electron diffusion coefficient in GaAs is $\mathcal{D}_n = 0.025$ m^2/s. So, the time taken will be $t \approx L^2/\mathcal{D}_n \approx 4\ 10^{-11}$ s. Four hundred-billionths of a second! The diffusion process in such tiny devices does not seem slow after all, does it?

In semiconductor devices of many types free carriers are distributed in the bulk of the device non-uniformly. The range of the density change is very wide: from a very small value up to a very large one, while the length L may be tenths or even hundredths of a micrometer. That's why diffusion processes play such an important role in the work of semiconductor devices.

The Path of the Drunk. To have a better understanding of the diffusion process let us deduce Eq. (24) in another way.

Imagine that we are able to watch the motion of a single molecule, such as the perfume molecule. Bumping into the air molecules it moves chaotically, every bump sending the unfortunate wanderer in a new direction (Fig. 26). Between the collisions the molecule covers the distance equal to the mean free path l. Find the distance L covered by the molecule from the starting point, after it has had N collisions.

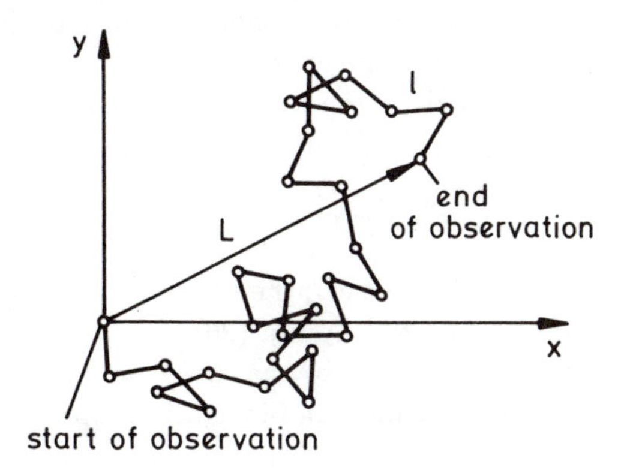

Fig. 26. As a result of N steps made at random, each step l long, the object covers the distance $L \approx l\sqrt{N}$.

This problem of random wanderings is known in mathematical statistics under the name of "the path of the drunk". For both the drunk and the molecule (though for different reasons) the direction of every next step is quite unpredicted and does not depend at all on the previous step. Therefore the problem we are interested in can be formulated in the following way: How far beyond the starting point will the drunk man have gone after he has made N steps, each step l long?

Let us draw in Fig. 26 two mutually perpendicular axes x and y, and let us characterize each step 1, 2, 3, ... i, ..., N by its projections on these axes $\Delta x_1, \Delta x_2, \Delta x_3 \ldots, \Delta x_i, \ldots \Delta x_N$; $\Delta y_1, \Delta y_2, \Delta y_3, \ldots \Delta y_i, \ldots \Delta y_N$. The values of Δx and Δy can, with equal probability, be either positive or negative. Their values lie within the limits from $-l$ to $+l$. Besides it is clear that $\Delta x_i^2 + \Delta y_i^2 = l^2$.

Let us calculate what the value L^2 will be equal to after N random steps:

$$L_N^2 = \left(\Delta x_1 + \Delta x_2 + \Delta x_3 + \ldots \Delta x_i + \ldots \Delta x_N\right)^2 + \left(\Delta y_1 + \Delta y_2 + \Delta y_3 + \ldots \Delta y_i + \ldots \Delta y_N\right)^2 . \qquad (25)$$

Then squared, the value L_N^2 will represent the sum of the squares of all the items: $\Delta x_1^2 + \Delta x_2^2 + \cdots + \Delta y_1^2 + \Delta y_2^2 + \ldots$ plus the sum of the following double products: $2\Delta x_1 \Delta x_2 + 2\Delta x_1 \Delta x_N + \cdots + 2\Delta y_1 \Delta y_2 + 2\Delta y_1 \Delta y_3 + \cdots + 2\Delta y_1 \Delta y_N + \ldots$.

We have already mentioned that while wandering at random, the values Δx_i and Δy_i can be with the same probability, either positive or negative. So when the number of steps is great, the sum of their double products is equal to zero. And every pair of the sum of the squares $\Delta x_i^2 + \Delta y_i^2$ is just equal to l^2. Therefore when N, the number of steps, is great

$$L_N^2 = Nl^2 \quad \text{or} \quad L_N = l\sqrt{N} \,. \tag{26}$$

So, having made N steps, l long each, the drunk (or the diffusion molecule) will cover the distance L, defined by Eq. (26).

But ... the number of "steps" of the molecule, atom, electron or the drunk is proportional to the time t and inversely proportional to the mean time between the collisions τ_0: $N \sim t/\tau_0$. The length of the free path of the particle l must be evidently equal to $l = v_T \cdot \tau_0$. Thus $L \sim v_T \cdot \tau_0 \sqrt{t/\tau_0} \sim \sqrt{v_T^2 \cdot \tau_0 \cdot t}$. Or considering Eq. (18) for the diffusion coefficient, $L \sim \sqrt{Dt}$.

So we have again come to Eq. (24).

Now, having received a general idea of diffusion, let us consider one of the most important diffusive processes in the physics of semiconductors, the diffusion of the minority carriers.

The Diffusion of Nonequilibrium Carriers. In some types of devices a certain type of diffusion, the diffusion of excess (nonequilibrium) carriers, plays the most important role. The peculiarity of this type of diffusion is that the excess carriers, while diffusing, move away from the place of their introduction into the semiconductor, and perish (recombine).

While studying in the previous section the processes of recombination, we established the following: if by illuminating a semiconductor, we create excess nonequilibrium carriers, then after the light has been switched off, the concentration of the excess carriers will diminish. Their lifetime τ, the time it takes the excess carriers to recombine, depends on the nature of the semiconductor, on the type and the density of the deep impurities in the material, and lies within the limits from $\sim 10^{-10}$ up to 10^{-2} s.

Let us discuss the experiment depicted in Fig. 27.

The semiconductor sample is illuminated by light with quanta of energy $h\vartheta$, which is greater than the energy of forming an electron-hole pair E_g.

The sample's surface is covered by a metal plate with a slit, cut in it. The light does not penetrate through the metal, so the electron-hole pairs created by light appear only directly under the slot (Fig. 27(a)).

We switch on the light and then switch it off at once (Fig. 27(b)). Let us assume the flash to be so short that the electrons which appeared under the action of light did not have time to either spread under the influence of diffusion or recombine. In this case the distribution of electrons immediately after the end of the light pulse will correspond to curve 1. What will happen to the nonequilibrium excess electrons next?

As we know, the nonequilibrium carriers will recombine (die out) with a characteristic lifetime τ. Besides, they will travel under the action of diffusion from the region of high density to the region of low density. Curves 2–4 (Fig. 27(b)) indicate what happens to the pack of the excess carriers due to the joint action of the processes of both, recombination and diffusion.

Electrons created by light, recombine during the time of the order τ. According to Eq. (24) while dying they have time to advance from the illuminated region where they have appeared to the shaded region covering the distance of the order of the diffusion length $L_{\mathcal{D}}$

$$L_{\mathcal{D}} = \sqrt{\mathcal{D}\tau}\ . \tag{27}$$

Let us make now the following experiment. We will switch on the light and leave it switched on for an indefinite time (Fig. 27(c)). Immediately after the light was switched on the nonequilibrium electrons existed only in the illuminated region (curve 1). Then the electron packet will begin to spread due to diffusion, though the process of generating carriers by light will dominate over the process of recombination. The concentration in the illuminated region will still increase (curves 2 and 3). Curve 4 corresponds to the stationary state. Electron distribution corresponding to curve 4 will remain stable while the light is switched on. The number of carriers created by light is exactly the same as the number of carriers perishing due to the recombination. At every point the number of carriers, delivered by diffusion from the neighboring region with a higher concentration, is equal to the sum of the number of carriers leaving this point for the neighboring region with a lower concentration and the number of the recombining carriers.

The nonequilibrium carriers, created in the illuminated region, penetrate into the shaded region as far as several diffusion lengths $L_{\mathcal{D}}$.

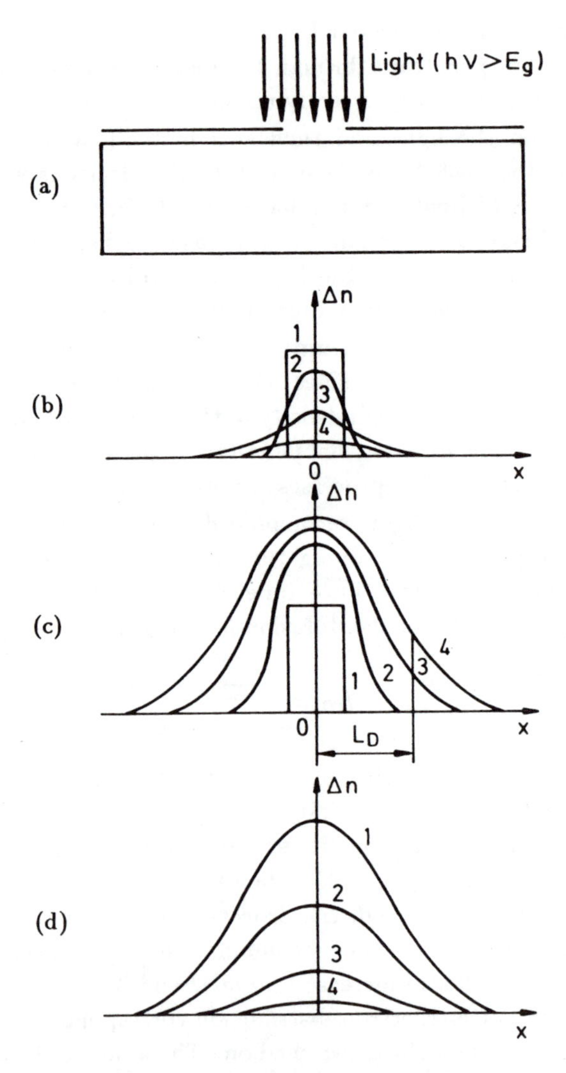

Fig. 27. The carriers which appeared in the illuminated part of the semiconductor propagate to the shaded region due to diffusion: (a) the specimen is covered by a metal plate with a slit in it; (b) the distribution of the excess carriers after a short flash of light, $(t_1 < t_2 < t_3 < t_4)$; (c) the excess carriers distribution at the moments of time $t_1 \div t_4$ $(t_1 < t_2 < t_3 < t_4)$ with the light switched on for a very long period of time; (d) the excess carriers distribution with the light switched off $(t_1 < t_2 < t_3 < t_4)$.

Figure 27(d) indicates what will happen if the light is switched off. Nonequilibrium carriers recombine, the recombination process taking time τ.

SUMMARY

Free carriers – electrons and holes – are constantly generated in the semiconductor under the action of thermal oscillations of the lattice. They also constantly recombine. So at a given temperature due to these two processes, generation and recombination, there is a certain equilibrium density of free carriers in a semiconductor. The value of the equilibrium concentration of electrons and holes in the semiconductor depends on the temperature and on both the type and the concentration of the impurities.

A certain exterior action, such as illumination, may increase the density of electrons and holes with regard to the equilibrium value. If any exterior action is eliminated, the concentration of the excess nonequilibrium carriers decreases, tending to approach the state of equilibrium. The process of the excessive carriers recombination is characterized by the carrier lifetime τ. The value τ decreases sharply when impurities causing the appearance of deep levels are present.

Deep levels increase both the recombination and generation rates.

Free carriers are in a state of random chaotic motion, which is very rapid. Under usual conditions the average velocity of such chaotic motion is $\sim 10^5$ m/s. The electric field causes a directed drift of carriers which is added to this chaotic motion. If the field is weak, the mean velocity of the drift is proportional to the electric field (Ohm's law). In a strong field the drift velocity of carriers is saturated. If the free carrier concentration in the semiconductor is nonuniform, then there appears diffusion, a flux of carriers directed from a higher density region to those regions where the density is lower.

Part III. PRINCES' PROFESSIONS (The Simplest Semiconductor Devices)

Business first, pleasure later.

Semiconductor devices, which we'll discuss in this part, are of a very simple design. They are just tiny semiconductor crystals with contacts. However, thanks to the wonderful physical properties of semiconductors even such simple devices can solve a lot of difficult, important and interesting problems in several branches of science and engineering.

Chapter 5. THERMISTORS

One must acquire the deepest knowledge of things to be able to discern the genuine relations between them, no matter how simple they might be.

G. K. Liechtenberg

A thermistor, or a temperature-sensitive resistor, is just a piece of semiconductor material with ohmic contacts – perhaps the simplest of the semiconductor devices.

But in reality, thermistors are used to measure and control temperatures ranging from 1 K up to the temperature of melted steel ($\sim$ 1800 K); to control the temperature of various elements of electrotechnical and radioelectronic machinery; to measure low pressures in vacuum equipment and also to measure the power of various microwave installations and the velocity of the motion of liquids and gases; to control the temperature of patients in hospital wards and to monitor at a distance the health of rare animals (Fig. 28). Besides all this they are used to study the spectra of the sun and stars. Thermistors are also used in fire alarm systems. They can be used as noncontact variable resistors, as time relays, as automatic potentiometers, as generators, amplifiers, low frequency modulators, as stabilizers and as safety devices.

Fig. 28. A telemetrical collar allows the tracking of rare animals and monitoring their health: body temperature is being measured by a tiny thermistor.

It's a good service record for such a small device, isn't it? And they have a long pedigree too – the first thermistors were manufactured as far back as 1890 by the great German physicist Nernst.

How does the single thermistor accomplish these various functions? Let us start with the simplest of them – measuring and controlling the temperature.

TEMPERATURE-SENSITIVE RESISTOR

The resistance of a semiconductor depends to a very great extent on temperature. A change in the thermistor resistance indicates a change in the ambient temperature. Everything is quite simple. But let us first discuss the problems which face the designers of thermistors. The first problem is which semiconductor material is to choose to manufacture the thermistor.

Low Temperatures

Suppose we are to measure low temperatures of the order of several Kelvin. We know that at absolute zero semiconductors simply become insulators. If the temperature, though above zero, is still very low, the semiconductor resistivity can be inadmissibly great. It is quite easy to establish it by means of Eq. (6) and the data given in Table 1. Even in InSb, which is a semiconductor for which the energy required to generate an electron-hole pair E_g is very small, the concentration of intrinsic electrons and holes $n_i = p_i$ with $T \sim 1$ K will make $\lesssim 10^{-400}$ cm^{-3}! In practice, however, the resistance of a sample, made of such material, will be defined not by the intrinsic conductance, (the latter being extremely small) but by the current leakage on the surface of the thermistor and also by the inevitable incidental impurities and so on. The resistances of different samples, though made under the same conditions, will differ from each other hundreds of times. No sample will have a stable resistance. In short, such thermistors will not be of much use.

We have already learned a lot about the semiconductor to deduce a way out. It is necessary to dope the semiconductor with a shallow impurity.

Provided the ionization energy ΔE of shallow impurity is small enough, and the density N large enough, the thermistor will have an acceptable resistance even if the temperatures are rather low (see Eq. (8)).

The necessary solution has been found. But is this answer exhaustive? No. We have now realized how to create a thermal resistance, sensitive in the range of low temperatures. We must however not forget about the phenomenon of impurity saturation. If we introduce into a semiconductor only one type of impurity whose ionization energy is very small, then the resistance will be sensitive to changes in temperature only within a very narrow temperature interval. In the region where the temperature is a bit higher, impurity saturation will occur (see Fig. 14) and the concentration of free carriers will no longer depend on temperature. But it is desirable, as a rule, to measure the temperature within a rather wide range.

What is there to be done then? The answer is that we must introduce into a semiconductor not just one type but many types of impurities with different activation energies ΔE. Knowing how sensitive semiconductors are to various impurities, we can easily estimate the practical difficulties of the precisely controlled, reproducible, multicomponential doping.

Nevertheless, modern technology can cope well with this complicated problem. In Fig. 29(a) we can see a tiny germanium thermistor for measurements in the temperature range of 1.3 up to 273 K.

The form, the sizes and the design of the frame of the thermistor can vary to a great extent. The silicon thermistors are shown in Fig. 29(b). They are used to measure and control temperature within the range of 10 to 400 K.

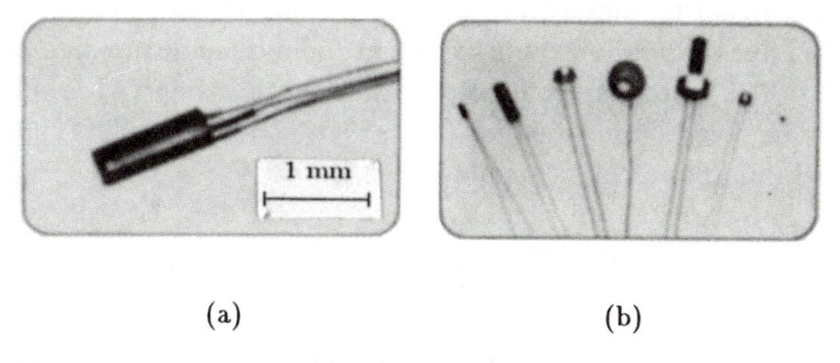

(a) (b)

Fig. 29. Thermistors. (a) Thermistor on the base of germanium with multiple doping. The working temperature is in the range of 1.3 to 273 K. Reproducibility is not worse than 0.0005 K. The thermosensitive element (a germanium film ~ 200 μm thick with a diameter ~ 0.6 mm) has a mass of only 0.0001 g. The tiny size of the device allows the measurement of temperatures in small volumes and guarantees a low thermal time constant of the device. (b) Different designs of silicon thermistors. (*Physics Today*, 1982).

High Temperatures

The semiconductor material, which high-temperature thermistors are made of, must fulfill a number of requirements. The first is quite clear.

The energy generating an electron-hole pair E_g in such material must be large.

In addition, when speaking of high temperatures, properties of the material such as its chemical and structural stability are of utmost importance. It is not easy to find a material able to survive multiple heating, say, up to 1000°C without changing its properties. Neither is it easy to select such contacts which can endure high temperatures and have the *temperature expansion coefficient* similar to that of the semiconductor itself, i.e. have the same size-temperature dependence. If that requirement is not answered, then, after a few cycles of heating and cooling, the contacts may either break away or cause the semiconductor to crack.

Of all the semiconductors mentioned by us, it is the semiconducting diamond and silicon carbide (SiC) that fulfill the requirements of possessing a high chemical and structural stability and a large value of E_g best of all. They are used to manufacture high temperature thermistors for temperatures of up to 300–400°C.

For still higher temperatures, up to 1200–1500°C, such materials are applied which due to their low conductivity at room temperature belong rather to dielectrics than to semiconductors: aluminium oxide (Al_2O_3), magnesium, zinc- and cadmium titanates ($MgOTiO_2$, $ZnOTiO_2$, $CdOTiO_2$), mixtures of titanium and cobalt oxides.

Neither Very High Nor Very Low Temperatures

Well, here at least things must be really simple! Scores (if not hundreds) of various semiconductors can claim to be the materials from which thermistors are to be made of.

One of the most difficult problems is how to select the best materials, which should be stable, reliable, simple and reproducible and also cheap (a very important condition for a device to be mass produced).

As a result of the long and hard fight, the royal throne in the kingdom of thermistors was taken by the princes belonging to the family of transition metals: manganese (Mn), cobalt (Co), nickel (Ni) and copper (Cu). Scores of types of thermistors are manufactured on the base of those materials. The majority of them are meant to be used within the temperature ranges of −60°C up to +120°C, though some of them can be used either at a lower temperature (up to $\sim$ 10 K) or at a much higher temperature (up to 300°C).

They may have various different forms and designs. Thermistors are manufactured as rods, tubes, disks, plates, beads and so on. The resistance value of various types of thermistors lies within the limits of several ohms up to several millions of ohms. (Even the term thermistor implies that its resistance is temperature-dependent. The resistance value is usually shown for a certain temperature close to the middle of the operating range. Very often it is +20°C.)

THE TEMPERATURE RESISTANCE COEFFICIENT

The temperature resistance coefficient is the most important parameter of the thermistor, regardless of its material, application or the temperature operating range.

The temperature resistance coefficient α_T shows the resistance-temperature dependence of the thermistor

$$\alpha_T = \frac{1}{R}\frac{\Delta R}{\Delta T} \,. \tag{28}$$

Equation (28) means the following: Let the thermistor resistance be equal to R at a certain temperature T. If there is a change of temperature equal to the value ΔT, there will be a change in the resistance too. It will then be equal to $R + \Delta R$. Thus, in accordance with Eq. (28), the temperature resistance coefficient represents a relative change in the resistance with the one degree change in temperature.

The resistance of the thermistor is defined by its geometrical sizes and by the resistivity of the material it is made of. The geometrical sizes hardly change with the change of temperature; the change in the resistance of the thermistor is defined by its resistivity-temperature dependence

$$\alpha_T = \frac{1}{\rho}\frac{\Delta \rho}{\Delta T} \,. \tag{29}$$

In metals the value α_T depends very little on temperature. It is positive (the resistivity of metals increases with the rise of temperature). For pure metals it is approximately 0.0035 K^{-1}. With the temperature change of 1 K the resistance increases about 0.35%.

The temperature operating range of the thermistor is always chosen in such a way that the temperature change causes an exponential change in the carrier density of the semiconductor that the thermistor is made of. The dependence of conductivity σ on temperature is then determined by an equation analogous to Eqs. (6), (8), and (11)

$$\sigma = \sigma_0 \, e^{-\frac{\Delta E}{2kT}} \,. \tag{30}$$

The resistivity being $\rho = \frac{1}{\sigma}$

$$\rho = \rho_0 \, e^{\frac{\Delta E}{2kT}} \,. \tag{31}$$

Knowing the dependence $\rho(T)$ defined by Eq. (31), we can easily find the value α_T

$$\alpha_T = \frac{1}{\rho} \frac{d\rho}{dT} = -\frac{\Delta E}{2k} \frac{1}{T^2} [K^{-1}] \,. \tag{32}$$

Thus, as it is shown in Eq. (32), in semiconductors the value α_T greatly depends on temperature ($\alpha_T \sim \frac{1}{T^2}$). It is negative. When heated the resistivity of the semiconductor decreases.

As for the value α_T, it is expedient to have it as large as possible. The greater the value α_T, the greater is the sensitivity of the thermistor resistance to the change in temperature, the easier it is to detect small changes of the temperature, and consequently, the more exact the temperature measurement and control will be. At first sight it might seem that it is quite easy to obtain very large values of α_T, especially at low temperatures, in accordance with Eq. (32). It is just necessary to choose a semiconductor whose activation energy ΔE is great enough, and... But we have become quite experienced by now not to let ourselves be deluded. If the relation $\Delta E/kT$ is too large – the carrier density, and consequently, the conductivity become inadmissibly small (see the section "Low Temperatures" in this chapter).

The value α_T for various types of thermistors lies practically within the range from -0.02 up to -0.08 K^{-1}. With a one degree change in temperature the resistance of the thermistor changes from 2 to 8%.

After we have discussed the main properties of the thermistors, let us consider their ability to cope with their numerous duties.

MEASURING THE TEMPERATURE

The temperature measurement is reduced to merely measuring the resistance of the thermistor. Knowing the resistance, by the equation analogous to Eq. (31),

$$R = R_0 e^{B/T} \tag{33}$$

we can determine the temperature; for the sake of practical calculations, however, it is reasonable to exclude from Eq. (33) the constant R_0.

Let us assume that at temperature T_1 the resistance of the thermistor is

$$R_1 = R_0 e^{B/T_1} \, . \tag{34}$$

At temperature T_2 the resistance will be equal to

$$R_2 = R_0 e^{B/T_2} \, . \tag{35}$$

Dividing (35) by (34), we obtain

$$R_2 = R_1 e^{B(\frac{1}{T_2} - \frac{1}{T_1})} \quad \text{or} \quad T_2 = \frac{BT_1}{B + T_1 \ln \frac{R_2}{R_1}} \, . \tag{36}$$

Knowing the value of the resistance R_1 at a certain temperature T_1, and the value of the constant B, by Eq. (36) we can determine the temperature, the resistance R_2 being known. The values of the constant B and the resistance R_1 at a certain temperature T_1 (which corresponds, as a rule, to the middle of the temperature operating range), are given in the certificate of the thermistor. The value of the constant B lies within the range of 25 K up to 25 000 K.

Besides, for thermistors, designed for most precise measurements, and for thermistors, designed to operate within a very wide temperature range, special graphs of resistance-temperature dependences can be applied. The procedure is very simple: after measuring the resistance of the thermistor, the temperature is to be read from the graph.

TEMPERATURE COMPENSATION

Many electromeasuring devices have to operate within a very wide temperature range. For instance voltmeters and ammeters often have to operate within the temperature range of −30 up to +50°C, without lowering the precision of their readings. The majority of voltmeters and ammeters have one of the main assemblies made of a light wire frame, rotating in the field of a constant magnet, with the current which is to be measured flowing through it. The frame is made of copper. The copper temperature resistance coefficient is $\alpha_T = +0.0039\ \mathrm{K}^{-1}$. Thus when the temperature change is 80 degrees, the resistance change of the frame is $\sim 30\%$!

It goes without saying that under such conditions the readings cannot be precise.

What must be done to improve the situation is the following:

A thermistor whose temperature-resistance coefficient is negative must be connected to the circuit in series with the metallic frame. Then the resistance of the frame will rise proportionally to the rise in temperature, while the resistance of the thermistor will drop, and the change in the resistance of the thermistor will compensate the temperature dependence of the resistance of copper.

But not everything is clear here. The temperature dependence of α_T for metals is very small, and that of the thermistor is very large. Besides, the values α_T are quite different for metals and for semiconductors. How can exact compensation be provided under such conditions?

The practical circuits include thermistors, connected to the high-stable manganin or constantan resistances both in parallel and in series. The parameters of the circuits are selected in such a way that they might provide compensation within a wide temperature range.

Thermistor temperature compensation lowers the sensitivity of the device parameters to the ambient temperature change dozens of times.

THE THERMISTOR AS A VARIABLE RESISTOR

All of us have to deal with variable resistances: they serve to regulate the volume in radios, compact disk players, TV sets and tape recorders. Every owner of these devices knows that after three or four years of operation, the rotation of the axis of a variable resistor (potentiometer) is accompanied by cracking and wheezing. Sometimes the volume may suddenly change without any visible reason or the sound disappears altogether. "Well, that's the end. The contact must have been worn out. It's time to change it." – the domestic expert makes the conclusion.

In a usual variable resistor, a metallic contact moves along the conductive layer and gradually rubs it off. The resistor fails altogether, and goes out of service though we turn the handle of the potentiometer only five to ten times a day.

Then there are circuits (for instance those of automatic control), in which it is necessary to change the magnitude of the resistance several times a minute. Then how long can an ordinary potentiometer last in such a circuit?

The so-called thermistors with indirect heating are indispensible in such circuits. The device represents a thermistor in the vicinity of which there is a tiny heating coil. With the current passing through it, it is heated, in turn heats the thermistor and changes its resistance. There are no mechanical contacts, or mechanical wear or any troubles connected with it.

The resistance of the heating coil is usually a few scores of ohms and the operating current is 20–40 mA. So, the power necessary to control the magnitude of the resistance is not large, of the order of a tenth or a hundredth of a watt. The tiny sizes of a thermistor and its heating coil make it possible to design thermistors with indirect heating which would have a very small thermal inertia: with the change of the current in the coil a new value of the resistance is established as quickly as in 5 to 10 seconds.

An important advantage of such thermistors over the ordinary potentiometers is their ability to control the resistance without any contact, at a distance, practically at any distance from the controlled equipment or circuit.

Thermistors with indirect heating are manufactured in large series, with various modifications, and are used in radiotechnical, electrotechnical and telemechanical devices.

They prove to be very effective for measuring the velocities of the motion of liquids or gases. The principle of measuring is very simple. The quicker the flow of the air around the thermistor (or the flow of the liquid, if the thermistor is plunged into it), the lower will be the temperature of the thermistor (and accordingly, the greater will be the resistance) provided the current in the heated coil is the same. If you don't believe it, lick your finger and blow on it!

The high temperature coefficient of the resistance makes it possible to use thermistors for measuring even small velocities of streams of liquids and gases up to ~ 1 mm/s.

BOLOMETERS AND SOMETHING ELSE

Bolometers are used to measure the energy of thermal radiation, usually very small. The source of this radiation is the light of the stars or of the sun, passing through a spectrometer and dispersing into thousands of rays, the energy of each of them very small.

The bolometer was invented in 1880 by the American astrophysicist, Samuel Lengly.

The semiconductor bolometer is in fact a thermistor but since it is intended to change its own resistance under the action of a very weak thermal flux, the design should answer quite a number of conditions.

The main element of the bolometer is a very thin (from fractions of a micrometer up to several micrometers) film of a semiconductor material applied on a glass or quartz substrate.

In order that the indication of the bolometer might not be affected by ambient temperatures, two similar devices are used, with identical characteristics, connected to a special circuit (the so-called "bridge circuit"*). In such a circuit the change of the resistance of one bolometer under the action of temperature fluctuations is practically fully compensated by exactly the same change of the resistance of the identical bolometer located nearby – the compensator.

By means of modern electronic circuits it becomes possible nowadays to detect the temperature changes of the bolometer as small as ten millionth of a degree. Thanks to that, semiconductor bolometers can detect the radiation whose power is ten millionth of a watt.

Now knowing the principle of the design of the thermistor we can understand how this device, simple as it is, can cope with its other numerous tasks, described at the beginning of this section.

In case of a fire, a sharp rise in the temperature causes an abrupt fall of the thermistor resistance. The relay, connected in series with the thermistor, will switch on the relay winding of a stronger wattage, and then depending on the circuit, there will be either the scream of a fire alarm siren, or a stream of foam from the automatically switched on fire extinguishing system to put out the fire.

Exactly the same system can be used for the distant control of the thermal regimes of various machines and mechanisms, for preventing the overheating of grain in elevators and also for the emergency call for a doctor, in cases where the patient's temperature is too high, etc.

Thermistors are used to regulate automatically the level of the liquid

---

*We will speak about this circuit in the next chapter where we will discuss the properties of resistance strain gauges.

in a vessel. Liquids, as a rule, conduct heat much better than air. This phenomenon is widely used for the automatic control of the liquid level. The current which passes through the thermistor heats it up to a certain temperature. While the thermistor is in the liquid (say, in a tank of water), its temperature corresponds to the given value. But if the water level is lowered, the thermistor appears to be in the air. The air does not conduct heat as well as the water. The same current begins heating the thermistor thereby raising its temperature.

Subsequently – everything goes as before. The resistance of the thermistor will drop abruptly. The relay will close the contacts, the actuator will be switched on and the tank will be filled with water. As soon as the water reaches the wanted level, the thermistor being in the water will cool down. Its resistance will grow, the relay will switch off the actuator and the stream of water into the tank will cease.

If a capacity is connected to a circuit containing a thermistor, then it is possible to design time-relays, pulse generators, modulators and many other schemes.

We will not consider them now. It is important to understand the physical phenomena on which the design and the application of thermistors are based. Then it would be easy to understand any other specific scheme.

Chapter 6. SEMICONDUCTOR STRAIN GAUGES

> There is a characteristic feature about people: before some wonderful discovery has been made, people doubt it can ever be made at all. And after it has been made, people are surprised it has not been made before.
>
> Jan Komensky

A semiconductor strain gauge is a semiconductor resistor whose resistance depends on the mechanical force compressing or stretching the resistor.

Unlike the thermistor which has a lot of different applications (we discussed them in the previous chapter), strain gauges are used almost exclusively to measure deformations and mechanical strains. The strain gauge does this type of work quite expertly within a wide range of values and under different conditions. Thanks to the strain gauge it became possible to measure the change in length within the limits of one hundred millionth of a per cent ($\sim 10^{-10}$) up to the value $\sim 5 \cdot 10^{-3}$. (A large majority of metals and alloys are destroyed by elongation about $5 \cdot 10^{-3}$ or 0.5%). Strain gauges can measure deformations which appear in ten millionths of a second, and they can detect deformations in steel and reinforced concrete constructions, the deformations, slowly accumulating for hours, days and months. By means of strain gauges we can detect and measure deformations both when the frost is severe (the lowest temperature recorded was ~ -100°C), and when the temperature is +300, +400°C, which is not infrequent during various technical processes.

THE NECESSITY OF MEASURING DEFORMATIONS

The range of deformations we come across in physics, engineering and biology is very wide. It can be as small as an atom or it can be so large that it can be seen with the naked eye.

Our brain is capable of perceiving the slightest shift of the sensitive elements of the ear as a sound – a shift as small as an atom.

On the other hand, the deformation of our muscles, governing our breathing and the movements of our hands and feet is as much as several centimeters.

In precise technical devices a deformation as large as 1 μm (10^{-6} m) may put the device out of order. On the other hand, the top of a skyscraper can vibrate because of strong gusts of wind, with the amplitude of vibration being of the order of a meter, with no dangerous consequences whatsoever.

The role of the relative deformation æ (i.e. the relation of the size change to the initial value) in physics and engineering is much greater than that of the absolute deformation. Speaking of the change in length, we have

$$æ = (l - l_0)/l_0 = \Delta l/l_0 \tag{37}$$

where l_0 is the initial length, and l is the final length.

The importance of the relative deformation is quite clear. It is proportional to the mechanical strength exerted on the construction or on some workpiece or on some part of a piece. In case the relative deformation exceeds the allowed value (which is quite definite for every material), it may result in a lot of trouble. Brittle materials crack and fail, plastic materials change their form irreversibly, eventually failing too.

When designing an aircraft, it is necessary to calculate the deformation under strain for every element and for every case: when taking off, when landing, when manoeuvring or when buffeted by gusts of wind. A modern liner or a supersonic fighter comprises a metal hull of the plane, an undercarriage, wings, turbines, screws, empennage – hundreds of workpieces. Which parts of the hull and of the wings are most "tense" under the strain? Where is the highest, perhaps the most dangerous, value of relative deformation? How are the forms and the designs of the workpieces to be changed so as to avoid any excessive strain and deformation? These questions are to be answered before beginning mass production in order to avoid possible accidents.

The turbine of a modern hydroelectric station represents an enormous cylinder whose mass is hundreds of tons, with blades (bulges of a most complex form) on the sides. A mighty stream of water beating on the blades, rotates the turbine, which causes the rotation of the power-station generator. One single blade being out of order results at best in the stoppage of the turbine and ceasing the delivery of energy to the consumers. At worst the broken blade might cause a bad accident with heavy losses in men and equipment. What strain does the blade experience? In what cross section are the strain and deformation especially great? It may be necessary to increase the thickness in one place and to change the shape in another?

Will the bridge be able to stand the impact of a bad flood and the blows of the hurricane simultaneously?

What deformations would appear in the construction in these three cases: under the usual, heightened, and extreme strain?

Without answering these questions it is impossible to design, create or test any modern technical equipment.

HOW IS DEFORMATION TO BE MEASURED?

Historically the very first way of determining deformation consisted in measuring the dimensional changes of a body which appear under the action of strain. It was done by means of a simple mechanical device, such as a ruler, a micrometer or callipers. This simple way of measuring deformations however had a lot of drawbacks. It was not very accurate; it could only be used to measure the static loads on the fixed stationary parts. This method was also quite inadequate in instances like measuring the almost inaccessible parts inside a mechanism. Last, but not least, this method only allowed the determining of the dimensional changes of a body as a whole, i.e. the average deformation. But the workpieces of a complicated shape can have in some of their parts strains and deformations exceeding the mean value by dozens of times.

MECHANICAL STRAIN SENSORS

Mechanical strain sensors or as they are sometimes called, extensometers, increase the precision of the deformation measurements.

The precision of measurements is increased on account of amplifying a small deformation by means of a system of levers. Ordinary scales are in fact a perfect extensometer. A small shift of about a few millimeters is transformed by a system of levers into a large swing of the scales pointer. The most perfect extensometers enable the measuring of the relative deformations of $\sim 10^{-5}$ and amplify small signals by a factor of $\sim (2 \div 3) \cdot 10^3$. The distance between the points at which the extensometer is fixed to the workpiece which is to be tested, the so-called base, is about one centimeter long in the case of mechanical tensometers. Measurements done by means of extensometers allow the determination of the average tension on the length as long as the base.

OPTICAL EXTENSOMETERS

These devices allow the diminishing of the base and increase the sensitivity. The system of mechanical levers, increasing small shifts, is substituted in optical extensometers by a system of mirrors. A beam of light is used as an indicator of the shifts (i.e. of deformation).

Good optical extensometers allow the measuring of the relative deformation $\sim 10^{-6}$ with the base of $\sim$ 5–6 mm. Since the small mirrors deflecting the light can be made very light, the optical extensometer can be applied to measure the variable deformations up to frequencies of 100–150 Hz.

The gain coefficient of the best optical extensometers is 30 000, and the base is only 1.5 – 2 mm. But these expensive, very complicated, unique devices can be used only in laboratories.

ELECTRIC STRAIN GAUGES

Electric strain gauges are based on changes which occur in their electric parameters under the action of the deformation of the workpiece which is being tested.

Capacitance, inductance and resistance can be used as such parameters.

Capacitive Strain Gauges

These strain gauges represent the capacitors, one plate of which is stationary. The other plate, the mobile one, is connected to the workpiece under test. The deformation of the latter results in the capacitance change of the capacitor. The strain gauge serves as the element of a high frequency generator, and the change in the strain gauge capacitance results in the change of the generated frequency. By the frequency change of the generator it is easy to judge the magnitude of the deformation in the place where the gauge is fixed.

Inductive Strain Gauges

These gauges are in fact coils with a mobile iron core. The core (or the coil) is connected to the workpiece to be tested. With deformation the core either moves in or moves out, the inductance changing accordingly.

Tensometers with electric strain gauges have a great advantage over any other kind of tensometers. They provide distant-reading, practically at any distance from the object under test.

But apart from advantages, both capacitive and inductive gauges also have a number of drawbacks.

Capacitive strain gauges are very sensitive to vibrations. It is very difficult to fix them to the workpiece which is to be tested and it may be quite impossible to test the piece whose configuration is complicated. Inductive strain gauges have the same disadvantages and besides, they have much greater weight and size.

THE WIRE RESISTANCE STRAIN GAUGES

The strain gauges of this type were proposed in 1938 independently and simultaneously by two American engineers – Simmons and Rudge. The principle of the work of these strain gauges is so simple that it just seems strange that the device was invented as late as that. Still more surprising is the fact that a few years after the first samples of strain gauges had appeared, hundreds of thousands of copies were manufactured which found a very wide application.

It worked this way: a very thin metal wire is glued to the workpiece which is being tested. If this workpiece compresses under the load, the wire compresses too. Its length l diminishes and its cross section increases. The ohmic resistance of the wire $R = \rho L/S$ decreases. With stretching, on the contrary, the length of the wire increases, and the cross section decreases. As a result, the resistance of the wire increases. With the change of the resistance it is possible to judge the amount and the sign of the deformation.

The resistance of the wire changes with the deformation almost instantaneously. Therefore the wire strain gauges can be used to measure deformation changing very rapidly.

The wire can be quite easily fixed (glued) to a detail of any configuration.

The wire strain gauge is as small as a postal stamp and its mass is fractions of a gram (Fig. 30). If necessary, such a strain gauge can be made as small as just fractions of a millimeter. The gauge is cheap, simple for production, reliable and stable.

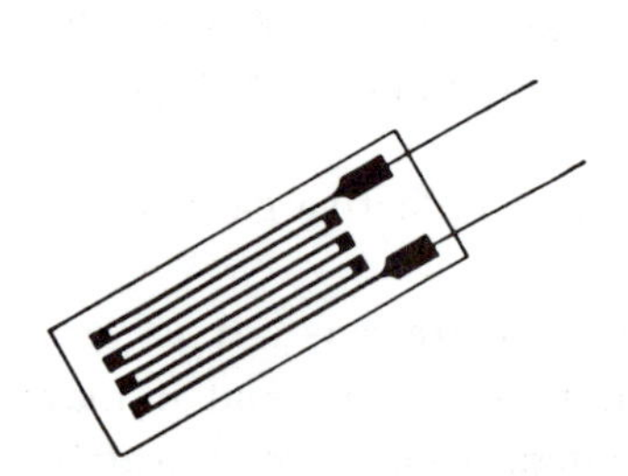

Fig. 30. A wire resistance strain gauge. To increase the resistance and sensitivity, it is made of several coils of wire. (C. C. Perry and H. R. Lissner, *The Strain Gauge Primer*, McGraw-Hill, 1955).

The fact that the resistance of a metal wire increases with stretching was reported as far back as in 1856 by William Thompson (Lord Kelvin), the creator of the absolute temperature scale at a meeting of the London Royal Society.

Why then did it take so much time between William Thompson's discovery and the invention made by Simmons and Rudge?

One of the reasons for that lies in the fact that the resistance change of the wire strain gauge at any real deformation does not exceed a tenth of 1 per cent. To detect such a change of resistance, to say nothing of measuring it, is not simple at all.

Let us obtain the expression connecting the magnitude of the relative deformation of the wire $æ = \Delta l / l_0$ with the relative change of its resistance $\Delta R / R_0$.

The initial resistance of the wire is

$$R_0 = \rho \frac{l_0}{S_0} . \tag{38}$$

Let us assume that it is only the length of the wire l that changes with deformation (stretching or compressing), and that the resistivity ρ and the cross section area S do not change at all. Then the wire resistance change ΔR_l with the change of the length l will make

$$\Delta R_l = \frac{dR_0}{dl_0} \Delta l = \frac{\rho}{S_0} \Delta l , \tag{39}$$

and the relative change $\Delta R / R_0$ will, evidently, be equal to

$$\frac{\Delta R_l}{R_0} = \frac{\Delta l}{l_0} = æ . \tag{40}$$

In reality, however, with the deformation, the cross section area changes as well. If we assume that the length of the wire and the wire resistivity are not affected by deformation and remain constant, then the change of the cross section will lead to the resistance change ΔR equal to

$$\Delta R_S = \frac{dR_0}{dS_0} \Delta S = -\rho \frac{l_0}{S_0^2} \Delta S . \tag{41}$$

(The minus sign in Eq. (41) reminds us of the fact that with compression the area S increases and the value R decreases.)

The relative change of the resistance on account of the change of the area will make

$$\frac{\Delta R_S}{R_0} = -\frac{\Delta S}{S_0} . \tag{42}$$

In reality both the wire length and the wire cross section change with deformation. Therefore the resultant change of the resistance will be equal to

$$\frac{\Delta R}{R_0} = \frac{\Delta R_l + \Delta R_S}{R_0} = \frac{\Delta l}{l_0} - \frac{\Delta S}{S_0} . \tag{43}$$

When using Eq. (43) one must bear in mind that the values Δl and ΔS have opposite signs. With compression for instance, l decreases and S increases.

It is also time to recall an important and interesting event. As far back as the beginning of the last century, the famous French physicist Siméon-Denis Poisson (1781–1840) established that with elastic deformations, while Hooke's law is valid, the relative change of the diameter of the wire $\Delta d/d_0$ is proportional to the relative change of its length

$$\frac{\Delta d}{d_0} = -\nu \frac{\Delta l}{l_0} = -\nu \, æ . \tag{44}$$

The coefficient ν is called the Poisson coefficient. Its value does not depend either on the sign of deformation (stretching or compressing), or on the magnitude of the deformation within the limits of elasticity. The value of the Poisson coefficient for metals lies within the limits from 0.25 up to 0.4.

It is easy to prove that within the relative change of the diameter of the wire $\Delta d/d_0$ the relative change of its cross section area will make $\Delta S/S_0 \approx 2\Delta d/d_0$.*

So, the relation between the relative change of resistance and the relative deformation will eventually be written like this

$$\frac{\Delta R}{R_0} = (1 + 2\nu) \frac{\Delta l}{l_0} = (1 + 2\nu) \, æ . \tag{45}$$

One of the main parameters of the strain gauge is the strain-sensitivity factor $\widetilde{F}$, defined as the relation of $\Delta R/R_0$ to the relative length change with the deformation, equal to æ. Equation (45) shows that for the metal

---

*Please do it.

wire strain gauge $\widetilde{F}$ is equal to

$$\widetilde{F} = \frac{\Delta R}{R_0 æ} = 1 + 2\nu \; . \tag{46}$$

It lies within the limits of $\sim 1.5 - 1.8$.[†]

The highest possible deformations æ which bring the metal constructions to the brink of failure, make, as it was mentioned above $\sim 0.2 - 0.5\%$. Thus, the highest possible resistance change of the wire strain gauge does not exceed the value of $\Delta R/R_0 = \widetilde{F}æ \sim 5 \cdot 10^{-3}$. And in many cases which are of great practical importance, the value $\Delta R/R_0$ is $\sim 10^{-4} - 10^{-5}$.

For measuring very rapid and minute changes of the resistance it is necessary to have very sensitive amplifiers. These amplifiers must also be very reliable, stable, comparatively cheap and available. Such amplifiers (operating on vacuum tubes) did not appear until the 1930s. Perhaps that could explain why Simmons and Rudge's discovery was made as late as that.

The wire strain gauges had faithfully served science and engineering for almost 20 years. But in the fifties there was a "second industrial revolution" in strain measurement engineering. This was caused by the appearance of semiconductor strain gauges.

SEMICONDUCTOR STRAIN GAUGES

In 1954 one of the leading journals of the world, *Physical Review*, published an article by the American researcher Charles Smith. He investigated the influence of the deformation on the resistance of germanium and silicon crystals. The results were striking.

The strain-sensitivity coefficient $\widetilde{F}$ in certain cases was over 100–150! Semiconductors demonstrated once again the gift of "the princess and the pea". The sensitivity to deformation was 100 times greater than the typical strain-sensitivity of metals.

But the surprises we have in store are not exhausted by just a large $\widetilde{F}$ value.

The value and even the sign of strain-sensitivity depended on the type of material subjected to deformation – whether it was of n- or of p-type. It

---

[†] In reality, however, the resistivity of metals ρ does change with deformation. It was shown as early as in 1881 by the well-known Russian physicist and educator, Orest Hvolson. But for practical use of the metal wire gauge, that change is negligible.

also depended on the level of doping of the material with either the donor or acceptor impurities.

The strain-sensitivity of germanium and silicon proved to be anisotropic to a very great extent. Its value depended on the crystallographical axis along which the samples were cut off from the germanium and silicon monocrystals.

Which of the Parameters Change Under Deformation?

The enormous value of strain-sensitivity, discovered by Smith, the possibility of the existence of both, positive and negative values* and the distinct anisotropy – all those factors show very clearly that the nature of the resistance change of germanium and silicon is quite different from that of metals.

Deformation, changing the geometrical dimensions of metals, affects their resistance change and is practically the main reason of this change. As we have seen, as a result of this mechanism of tensosensitivity, the value of the coefficient $\widetilde{F}$ is about $1 \div 2$.

The value $\widetilde{F} \gtrsim 100$ cannot be explained by the change of geometrical dimensions. Besides, the latter always results in the positive value of $\widetilde{F}$; the values Δl and $\widetilde{F}$ have the same sign. In silicon and germanium the values $\widetilde{F} \approx -100$ have been observed! The resistance of the samples decreases very much with stretching and increases with compression.

So? That means that (see Eq. (38) the resistivity of the material ρ changes greatly with the deformation of the semiconductor crystals.

The value ρ is determined by the concentration of the free carriers in the material and by their mobility. Which of them changes under the action of deformation?

Speaking generally, both of them can change – mobility and concentration.

The deformation changes the distance between the atoms of the crystal lattice of the material and consequently, the forces of their interaction. The latter results in the energy change of the electron-hole pair generation E_g. In the impurity semiconductor the deformation results in the change of the distance between the impurity atom and the atoms of the lattice

---

*In accordance with Eq. (46) strain-sensitivity factor $\widetilde{F}$ is positive when the resistance grows with stretching and falls with compression. (ΔR and Δl have the same sign).

surrounding it, the ionization energy ΔE changing too. Therefore, the free carrier density in both, the impurity- and intrinsic semiconductor can be said to be deformation-dependent.

But under the conditions of the experiments made by Smith, the free carrier concentration change was quite negligible.

What played the main role in the effects observed by him was the mobility change of the free carriers – electrons and holes. *The carrier effective mass m^* changed under deformation.*

It is interesting to establish the relation between the change of the effective mass m^* and the effects observed by Smith. When the value of the relative deformation æ of silicon reaches the value $\sim (2 \div 4)\ 10^{-3}$, silicon when being stretched, breaks up. In accordance with Eq. (46) when $\widetilde{F} \approx 100$, the relative change $\Delta R/R_0$ and consequently, the relative change of the effective mass of the carriers m^* is then 20–40%.

The Main Property of Crystals

The crystal structure of the semiconductor materials plays the main role in the analysis of strain-sensitivity.

Anisotropy, i.e. the dependence of the physical properties on the direction in a crystal, is known to be the main property of crystals.

In the experiments made by Smith, anisotropy is displayed quite distinctly. Depending on the crystallographic axis along which the deformation force is exerted, the value $\widetilde{F}$ for the silicon samples lies within the limits from -100 up to $+200$, provided the concentration of the doping impurity, the temperature and other parameters of the samples are the same.

The term "crystallographic axis" is explained in Fig. 31, where an elementary cell of the silicon crystal is shown. The silicon monocrystal consists of numerous tiny cells, absolutely identical, the sizes of which are about $5 \cdot 10^{-10}$ m. Though the cell is called elementary, the spatial distribution of atoms inside the cell is seen to be rather complicated. Eight atoms (in Fig. 31 they are marked by numbers 1–8) form a cube. There is also an atom in the centre of each side of the cube (9–14), four more atoms (15–18) are located inside the elementary cell, on the spatial diagonals of the cube, linking atoms 1 and 7, 2 and 8, 3 and 5, 4 and 6.

Now let us imagine that the heroine of the book *Alice's Adventures in Wonderland* by Lewis Carroll, who managed to become so small that she could get into a rabbit's hole, succeeded in becoming still smaller, say a billion times smaller. Being as small as an atom, Alice now begins her travels inside the silicon monocrystal.

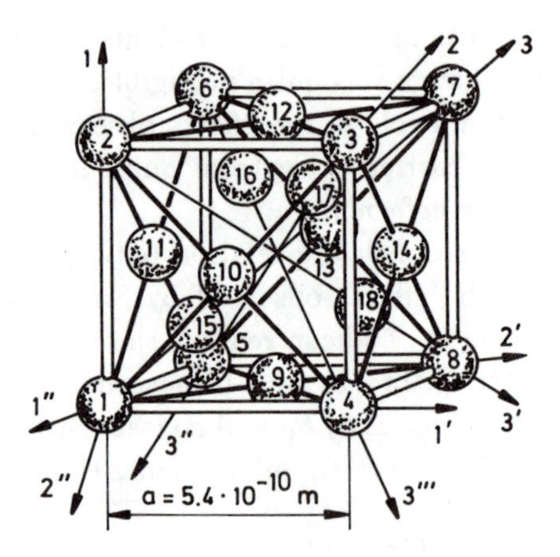

Fig. 31. An elementary cell of a silicon crystal. Elementary cells of germanium, gallium arsenide, diamond, one of the polytypes of silicon carbide and some other semiconductors have the same form.

The traveller could take various routes. She could move from atom 1 to atom 2 and then continue moving in that direction as far as she wanted (route 1). Due to the strict spacing of the crystal lattice, the landscapes that she would view would be repeated again and again. Alice would meet the silicon atoms quite regularly, each time at the distance of a_0, equal to the lattice constant. If in addition to her magic height, Alice could also have magic sight and could see the electrons, rotating in their orbits, she would see that the configuration of the electron orbits linking the atoms were also absolutely the same and that they were repeated at the same distances equal to the value a_0.

From Fig. 31 it is clear that had Alice chosen route $1'$ instead of route 1, from atom 1 to atom 4 (or route $1''$ from atom 5 to atom 1), she would have seen the same landscapes as on route 1. The routes (or using the scientific terms, the crystallographic directions) 1, $1'$ and $1''$ are absolutely equivalent.

Crystallographic directions 2, $2'$ and $2''$, crossing diagonally the sides of the cube are quite equivalent to each other, but they are quite different from directions $1-1''$. In directions $2-2''$ the distances between the neighboring atoms are equal to $a_0\sqrt{2}$, and not to a_0. The electron orbit configurations linking the atoms, will also be quite different.

Now let us leave Alice in her wonderland and return to the free carriers, electrons and holes, making their continuous travels in the cubic lattice of the crystal. As we remember, the free carrier effective mass m^* is defined by the character of the periodic electric fields of the lattice atoms which the carrier passes by (Chapter 4).

Considering the properties of the elementary cell, we see that the structure of those fields in various crystallographic directions is different. Hence the values of the effective mass of electrons and holes along various crystallographic axes may also be different. Even more different can be the changes of the effective mass under the action of the deformation directed along various crystallographic axes of the crystal. Just taking a look at Fig. 31 we may see that the stretch directed, say, along axis 1, will deform the crystal in a way quite different from that made by the stretch directed along the diagonals of the sides of the cube (axes $2-2''$) or along the spatial diagonals (axes $3-3'''$).

It is just this anisotropy in the change of the effective mass of the carriers that explains the effects observed by Smith. In *p*-type silicon, say with the stretch of the crystal along the spatial diagonal of the cube, depending on the doping impurity concentration, $\widetilde{F}$ can lie within the limits from +80 to +200. With the same stretch along the edge of the cube, the strain-sensitivity coefficient is close to unity; with the deformation along that axis, the hole effective mass practically not changing at all. In *n*-type silicon, on the contrary, the electron effective mass changes most of all, with the deformation of the crystal directed along the edge of the cube. With the deformation in that crystallographic direction, $\widetilde{F}$ lies within the limits from −60 to −90, depending on the level of doping. While with the deformation along the spatial diagonal of the cube, the change of the electron effective mass is much weaker ($\widetilde{F} \approx -5$).

From Physical Investigation to Practical Devices

Though scores of years have passed since the work by Charles Smith was published, it is still referred to very often. It has become a classical piece of work and has greatly helped in the studying of some subtle peculiarities of the motion of electrons and holes in semiconductors. Besides, the data obtained by Smith almost immediately resulted in a wide range of practical application.

An enormous strain-sensitivity of silicon and germanium, discovered by Smith, enabled the engineers to change the wire resistance strain gauges for the semiconductor strain gauges whose sensitivity is 100 times greater!

There were certainly a lot of difficulties to be overcome on the way from a physical research to a practical device. But the hundredfold increase in sensitivity was worth trying it.

The first obvious difficulty lay in the fact that silicon whose properties make it the most suitable material to produce strain gauges is very brittle. Meanwhile the ability to bend the strain gauge is one of the main advantages of the wire strain gauges. It is this property which makes it possible to research the deformations in the workpieces of very complicated configuration.

The thinner the silicon strip, the smaller the radius of the ring into which it can be bent without making cracks on its surface. The idea was to find a way to manufacture very thin silicon strips. It is seen from Fig. 32 that engineers have coped with this difficult technical problem.

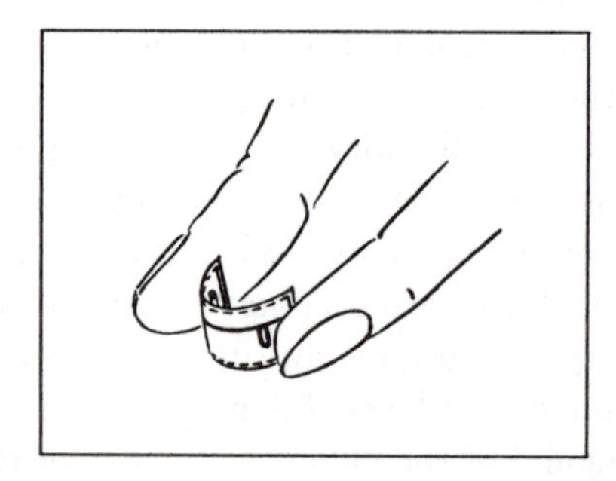

Fig. 32. At the end of the 1950s it was possible to manufacture the silicon strain gauges several dozens of microns thick. Such a gauge can be folded into a ring with a diameter of less than a centimeter.

The second difficulty to overcome was referred to as "the great temperature sensitivity" of the semiconductor resistance. Just fancy that! It is that very property which makes semiconductors wonderful temperature-sensitive resistors!

Using resistance strain gauges, we judge the amount of deformation by the resistance change. But suppose there is no deformation, and still the temperature of the strain gauge has changed. The measuring device will register the resistance change; there will be a signal of the so-called "seeming deformation".

It was clear from the very beginning of course, that in order to produce semiconductor strain gauges it is necessary to use semiconductors at temperatures corresponding to the impurity saturation (see the section "Donor Impurity" in Chapter 3). Then the resistivity-temperature dependence is rather weak and can be determined only by the semiconductor mobility change with the change of temperature (Chapter 4). But even in the impurity saturation region the temperature coefficient of the resistance α_T of semiconductors is much higher than that of the special highly stable metallic alloys which were used to produce the wire strain gauges.

There is a wonderful circuit named after the English physicist Charles Witston, who was the first to use it. It is a very simple circuit, the so-called "Witston bridge" which has served science and engineering faithfully for 150 years. It allows both to fully compensate the false signal caused by the heating of the strain gauge and to increase the sensitivity of the strain gauge schemes.

In the previous section, when describing the work of bolometers, we mentioned the bridge circuit. Let us now discuss in detail the properties of this wonderful circuit.

It is seen from Fig. 33(a) that the circuit is really simple. It contains only four resistances, connected in two parallel branches, each branch containing two resistances connected in series.

Let us first solve a very simple problem: let us calculate the potential difference between points A and B (Fig. 33(a)).

The current I_1 flowing in the branch, containing the resistances R_1 and R_4 is, evidently, equal to $I_1 = V_0/(R_1 + R_4)$. And the voltage drop at the resistance R_4 is equal to $V_4 = V_0 \cdot R_4/(R_1 + R_4)$. Analogically, the voltage drop at the resistance R_3 is equal to $V_3 = V_0 R_3/(R_2 + R_3)$. Consequently, the potential difference between the points A and B is equal to $V_4 - V_3$

$$V_4 - V_3 = V_0 \left[\frac{R_2 R_4 - R_1 R_3}{(R_1 + R_4)(R_2 + R_3)} \right] .$$

Provided the following condition is fulfilled

$$R_2 R_4 = R_1 R_3 \tag{47}$$

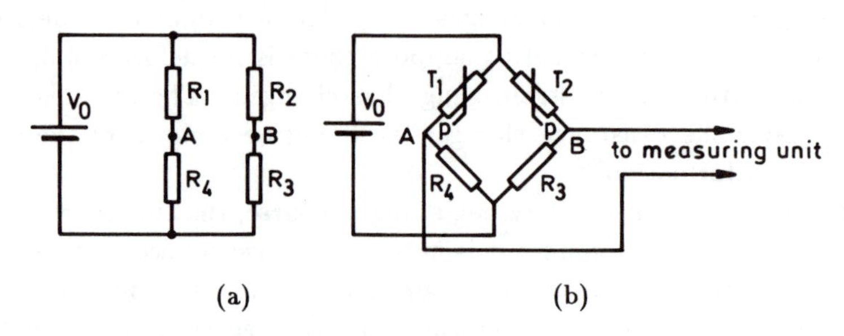

(a) (b)

Fig. 33. Witston Bridge. (a) With four resistors ($R_1 - R_4$). (b) With two strain gauges (T_1 and T_2) and two resistors (R_3 and R_4).

the potential difference between the points A and B is equal to zero for arbitrary voltage V_0 and values of the resistances R_1, R_2, R_3 and R_4. If Eq. (47) is fulfilled, the Witston bridge *is balanced.*

Figure 33(b) shows the circuit of the Witston bridge which is very often employed in practice. The circuit contains two strain gauges. One of them is fixed as usual (most often by means of a special glue to the workpiece which is to be tested). The second strain gauge (which is often called "idle" or "dummy") is placed close to the first one, but it is not fixed to the device under test. Or, if it is fixed, it is fixed in such a way that the deformation of the workpiece might not lead to the deformation of the auxiliary strain gauge (for instance through an elastic gasket).

Strain gauges T_1 and T_2 are chosen to be as identical as possible. Modern technology allows to manufacture strain gauges with quite well reproducible parameters. Thus, not only the initial resistances of the strain gauges R_1 and R_2 are equal, but also any other resistance changes caused by heating or say, by illuminating, are equal too. The resistance values R_3 and R_4 are also chosen to be equal. In the initial state the bridge is balanced.

Then no matter how much the strain gauge resistances change under the influence of the temperature or of the illumination, the values R_1 and R_2 remain equal to each other. Consequently, Eq. (47) remains valid. The bridge remains balanced and there is no voltage at the input of the measuring device. There will be no potential difference between the points A and B unless the workpiece and the operating strain gauge are subjected to deformation.

Using the auxiliary strain gauge allows for compensating the hindering signals of the seeming deformation.

The joint efforts of physicists, technologists and engineers have helped to remove the obstacles preventing a wide application of semiconductor strain gauges. In 1957, the first experimental specimens were produced and mass production of resistance strain gauges began in 1959.

What Strain Gauges Can Do

First of all they can measure deformations and strains in the workpieces of machines and in various constructions. Without diminutive – millimeters and fractions of millimeters long – semiconductor strain gauges (Fig. 34) it is impossible to create gigantic dams, antisill barrages weighing millions of tons, hulls of ships able to withstand a storm and hulls of planes which can be subjected to hailstorms and to the mad gusts of wind, the springs of multi-ton trucks and the antiseismic foundations of skyscrapers. When elaborating and designing such structures, semiconductor strain gauges enable us to measure the strains which appear in the sites and workpieces and to choose the optimal form, materials and the necessary safety margin.

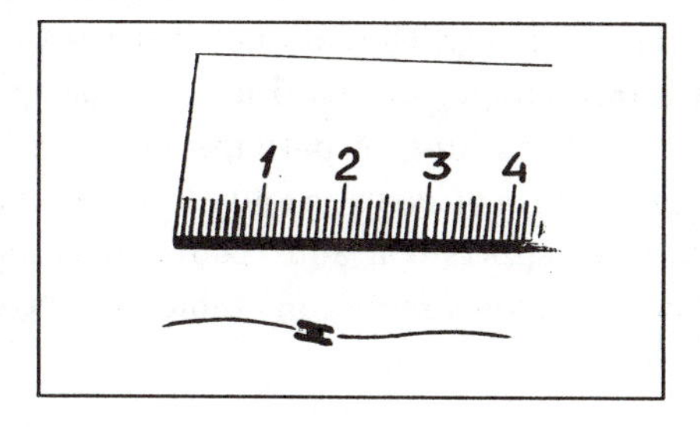

Fig. 34. The size of the semiconductor strain gauge may be a few millimeters or even fractions of a millimeter.

Besides this application, "according to the speciality", the semiconductor strain gauges are also used for measuring certain very important parameters, both physical and biological. Among the most important devices, made on the base of semiconductor strain gauges, are accelerometers, pressure transducers, and also medical and biological instrumentation.

Accelerometers, devices measuring acceleration, are the most essential part of autopilots, of missile guidance systems and of other equipment.

An accelerometer on the base of a semiconductor strain gauge represents a light, diminutive ($l \sim 1$ cm) deformable beam, at the tip of which a load of a certain exactly known mass m is fixed. The force f deforming the beam is proportional to the acceleration a ($f = ma$), and the readings of the strain gauge are known to be proportional to the deforming force f. Therefore the readings of the strain gauge can be graduated in units of acceleration. Depending on the construction, the sensitivity of the accelerometer can lie within the limits from $\sim 10^{-2}$ to 10^2 ohm/g. And the whole scale of the device is (10–1000) g.*

The design of pressure transducers is very simple. A resistance strain gauge is fixed to a light diaphragm whose deformation, measured by the strain gauge is proportional to the pressure. Such transducers, diminutive, resistant to vibration and to other hindering factors, proved to be very convenient when working at the experimental stands, in vacuum sets, and when measuring pressure in the flux of liquid or gas.

Explaining to Faust why he had to sign the oath to sell his soul to the Devil with his blood, Mephistopheles exclaimed, "Blood is a special kind of liquid!" Nevertheless, semiconductor strain gauges are widely used to measure blood pressure.

Catheters with semiconductor strain gauges are introduced directly into the blood vessels. Being so tiny, they do not disturb the blood flow.

Semiconductor strain gauges are used in monitoring systems to control the volume and rate of breathing of patients both on land and in cosmic space, to monitor the rhythm of systoles and blood pressure in blood vessels; to monitor the efforts of sportsmen and cosmonauts made under extreme conditions, and also to conduct numerous biological investigations.

Chapter 7. PHOTOCONDUCTORS

In nuances lies the truth.

A photoconductor is a semiconductor specimen with ohmic contacts whose resistance changes under the action of light.

---

*The somewhat unusual units of measuring (g) is often used in engineering when measuring accelerations. The letter (g) is used to denote the acceleration of the free fall of the body, g = 9.8 m/s². The sensitivity 10 ohm/g means that under the action of the acceleration of 1 g the strain gauge resistance changes to 10 ohms.

THE DIFFICULTY OF BEING EXACT

The definition of the photoconductor is simple and understandable. It is analogous to other semiconductor devices known to us: the thermistor and the strain gauge. Nevertheless an attentive reader has every right to be displeased.

When discussing bolometers, we described them as semiconductor specimens whose resistance changes under the action of light including the luminous radiation of the sun and the stars. What then is the difference between the photoconductor and the bolometer?

THE IMPORTANCE OF BEING EXACT

The physical phenomena, on which the operation of the photoconductor and the bolometer are based are absolutely different. Though on the face of it they seem very alike; in both cases the resistance of the semiconductor specimen falls under the action of light.

To understand the difference between the bolometer and the photoconductor let us conduct a simple experiment (Fig. 35).

A silicon plate fitted with ohmic contacts is connected in series with the load resistor R_l and the battery V_0. The plate can be illuminated by a source of light. By means of filters (or of a spectrograph) a radiation with a desired wavelength λ is separated from the spectrum of the source of light. A modulator is placed between the source of light and the plate. When the modulator is opened the light falls on the plate. When the modulator is closed the plate is darkened.

Let the modulator first be closed so that the dark current determined by the dark resistance of the plate flows through the plate and through the load resistor R_l. At the moment $t = 0$ the modulator opens and the plate becomes illuminated. The resistance of the plate falls, the current in the circuit increases, and the voltage drop on the load resistor R_l increases too, proportionally to the current. The voltage change on the load resistor is registered on the screen of the cathode-ray oscillograph.

Preparations over, we now begin the experiments. First we choose a set of filters transmitting infrared radiation with a wavelength of $\lambda \sim 5$ μm. We then open the modulator and some time later, at moment t_0 close it. On the screen of the oscillograph there is a curve 1, reflecting the change in time of the voltage drop V_l on the resistor R_l. When the modulator

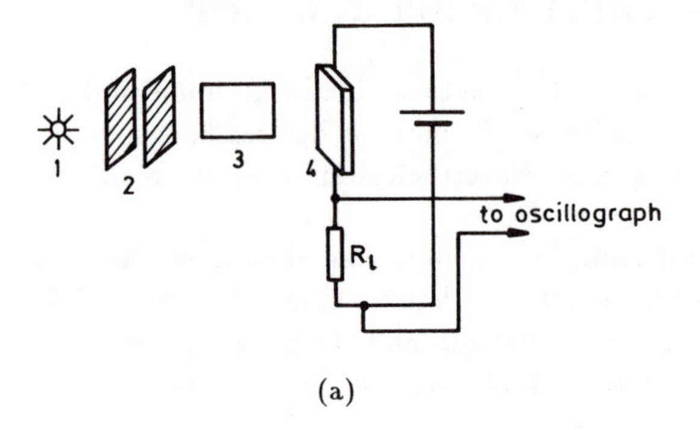

(a)

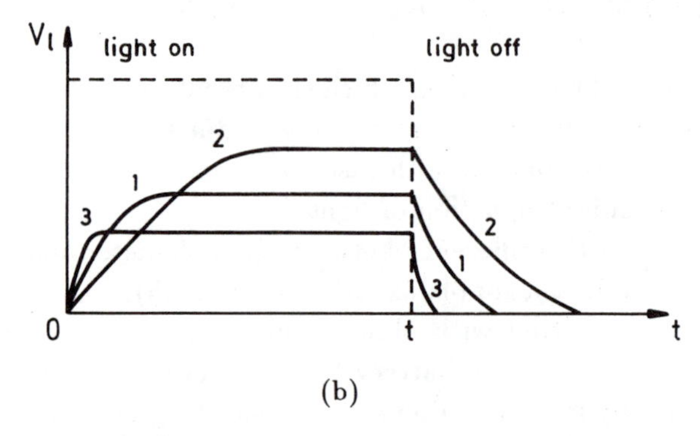

(b)

Fig. 35. An experiment demonstrating the difference between the bolometrical and photoconductivity modes of work. (a) 1 – the source of light, 2 – the filters, 3 – the modulator, 4 – the silicon plate with ohmic contacts. (b) The voltage on the load resistance as a function of time. Curves 1 and 2: $\lambda \approx 5$ μm, curve 3: $\lambda \approx 1$ μm. Curve 1 – the silicon plate in air. Curve 2 – the silicon plate in the vacuum tube.

opens, the resistance of the plate begins to diminish, the current in the circuit increases, so does the voltage V_l. Some time later a stationary state is reached; the resistance of the plate remains stable. After the modulator is closed at the moment t_0, the resistance of the plate begins to increase, and the voltage V_0 begins to fall.

Now, without changing anything in the scheme of the experiment, we will place a silicon plate into a vacuum vessel and open the modulator again. The $V_l(t)$ dependence will change quite visibly (curve 2). It is clear that the resistance of the plate now changes more slowly: the rise time and the fall time of the current have become appreciably longer. Besides, under the action of the same beam of light the resistance of the plate falls lower than when the experiment took place in the air.

Now let us change the set of filters in such a way that the wavelength of light falling onto the specimen λ should be $\sim 1\ \mu$m. Let us repeat these experiments in the air and in the vacuum. In both cases the screen of the oscillograph shows the same curve 3. Placing the silicon plate into the vacuum tube, λ being equal to $\sim 1\ \mu$m, will affect neither the periods of the rise or drop of the current, nor the magnitude of the stationary signal.

Let us discuss the results of the experiment.

The wavelength $\lambda \sim 5\ \mu$m corresponds to the energy of a photon $E_{ph} = hc/\lambda \approx 0.25$ eV. This value is smaller than the energy of the electron-hole pair generation in silicon $E_g \approx 1.1$ eV. Hence the light whose wavelength is $\lambda \approx 5\ \mu$m cannot generate electrons or holes. Nevertheless, falling onto the silicon crystal, the light with this wavelength is absorbed. It is absorbed by the free carriers of the crystal (electrons or holes). The photon energy passes over to the free electron (or the hole) and then, as a result of their collisions with the lattice, it passes over to the silicon lattice. The crystal is heated. On account of the temperature rise of the crystal lattice, the carrier concentration increases and the oscillograph registers the increase in current in the circuit.

The silicon plate operates in the bolometer mode: the conductance change is conditioned by the temperature change of the crystal.

It is clear that should the conditions of the heat sink change, the thermal condition of the bolometer will change too. This effect is observed when the specimen is placed into a vacuum tube. The heat sink becomes worse: now the specimen can lose heat only by means of radiation. Cooling on account of air convection is excluded. So on one hand heat equilibrium is reached more slowly. On the other hand, the luminous flux incident on the plate being the same, the temperature will be higher than when operating in the air, hence the resistance change will be greater. These features can be clearly seen when comparing curves 1 and 2 (Fig. 35).

The photon energy $E_{ph} \approx 1.25$ eV corresponds to the wavelength $\lambda = 1\ \mu$m. In the given case $E_{ph} > E_g$. The absorption of such light quantum by the silicon crystal creates an electron-hole pair.

The increase of the conductivity under the action of light conditioned not by the temperature change in the crystal, but by the increase of the number of free carriers, is called ***photoconductivity.*** The free carriers appear as a result of the absorption of light quanta. In the given example the specimen operates in the mode of a photoconductor. It is clear that the change of the heat sink does not affect the photoconductivity parameters: both in the air and in the vacuum vessel the conditions for the operation of the photoconductor are the same.

The absorption of the quanta, accompanied by the creation of the free carriers: an electron, a hole or an electron-hole pair, is called ***intrinsic photoeffect.***

Now we can give an exact definition: a photoconductor is a semiconductor resistor, whose resistance changes under the action of light on account of the intrinsic photoeffect.

THE INTRINSIC PHOTOEFFECT

Without being mentioned, this effect occurred in our narration quite a number of times. First in Chapter 1, when the question of the breaking of electron bonds creating electron-hole pairs was discussed. Then in Chapter 4, where the problem of exposing deep impurities as killers was considered and also when discussing the problem of the minority carrier diffusion.

In this section we will discuss the characteristics of the intrinsic photoeffect determining the main parameters of the photoconductors.

The Absorption Coefficient

Let a beam of light, incident on the surface of the specimen, have at $x = 0$, N_0 quanta falling per unit of the area per second (Fig. 36). Penetrating inside the semiconductor, the number of photons will be diminishing on account of the absorption. The absorption coefficient α serves as a quantitative measure of this effect. The greater the value of α, the faster diminishes the intensity of the luminous flux from the surface to the depth of the material.

The photon density diminishes along the coordinate x according to the following law

$$N(x) = N_0 \, e^{-\alpha x} \; . \tag{48}$$

The photon flux density drops exponentially from the surface into the semiconductor.

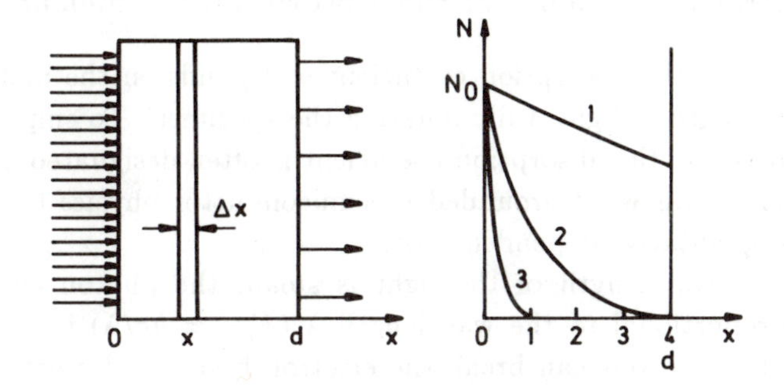

Fig. 36. The definition of the absorption coefficient. (a) The density of light flux N drops exponentially from the surface into the semiconductor in accordance with Eq. (48). (b) Three dependences $N(x)$ corresponding to the three values αd. Curve 1 - $\alpha d = 0.1$; 2 - $\alpha d = 1$; 3 - $\alpha d = 5$.

It is clear from Eq. (48) that the dimension of α is $[m^{-1}]$ (in books and papers on the physics of semiconductors α is more often measured in cm^{-1}.)

The character of the photon density distribution in the specimen depends on α and on the thickness of the specimen d.

In accordance with Eq. (48) the photon density at the back side of the specimen whose thickness is d is equal to $N_1 = N_0 \exp(-\alpha d)$.

The photons which have reached the back side of the plate had avoided absorption. The number of photons absorbed in the specimen (per unit area) is the following

$$N_2 = N_0 - N_1 = N_0 - N_0 e^{-\alpha d} = N_0(1 - e^{-\alpha d}) . \tag{49}$$

If $\alpha d \ll 1$, i.e. if either the coefficient of absorption is very small or the specimen is very thin, then $e^{-\alpha d} \approx 1$ and with $x = d$ (at the back side of the specimen) $N_1 \approx N_0$. The distribution of the photon flux in the specimen is practically uniform. Almost all the photons which got into the specimen, fly out of it, unabsorbed.

With $\alpha d \approx 1$ the photon density at the back side is approximately e times lower ($e = 2.72$) than on the front side of the specimen. An appreciable part of the luminous flux ($\sim 63\%$) is absorbed in the specimen.

With $\alpha d \gg 1$ the luminous flux is fully absorbed in the specimen. Moreover, practically all the photons are absorbed in the narrow region next to the surface (Fig. 36, curve 3).

The Spectral Dependence of the Absorption Coefficient

The value of the absorption coefficient α depends on the material and on the wavelength of light λ illuminating the specimen. To emphasize this last circumstance, the absorption coefficient is often designated as $\alpha(\lambda)$.

Now being somewhat grounded in semiconductor physics let us try to predict the qualitative dependence $\alpha(\lambda)$.

When the wavelength of the light is small, the photon energy E_{ph}, inversely proportional to the wavelength λ ($E_{ph} = hc/\lambda$) is great. With $E_{ph} > E_g$ the photons can break the electron bonds and create electron-hole pairs. Since the atoms of the lattice are numerous ($\sim 10^{22}$ cm^{-3}), the absorption in this wavelength range is expected to be large. With the increase of the wavelength of light, and consequently, the decrease of the photon energy to the value $E_{ph} \sim E_g$ the coefficient of absorption will drop.

The so-called "red boundary" of the intrinsic photoeffect corresponds to the energy $E_{ph} = E_g$. What is the meaning of this term?

The photoeffect is called *intrinsic* when $E_{ph} > E_g$. It is so-called because if $E_{ph} > E_g$ electrons and holes appear in pairs, exactly the way they do in the intrinsic semiconductor under the action of the thermal vibrations of the lattice. It goes without saying that the intrinsic photoeffect disappears when $E_{ph} < E_g$.

As for the term "red", there is no political shade of meaning here. The photons possessing the smallest energy visible to the naked eye correspond to the red light. (The photons possessing still smaller energy correspond to the infrared light.) When the photon energies of two sources are compared, the physicist can state quite definitely that the one whose quantum energy is smaller, is "redder" than the other. Even if both sources emit X-rays, for which, strictly speaking, the idea of color makes no sense. If due to some effect, the photon energy decreases, the light is said to "redden". It is slang, of course, but it is so widely used that specialists stopped noticing it.

The photons whose energy is $E_{ph} = E_g$ are the "reddest" photons of those able to cause the intrinsic photoeffect.

Photons whose energy is $E_{ph} < E_g$ cannot generate electron-hole pairs. But impurities are always known to be present in semiconductors. Even if $E_{ph} < E_g$, the energy of a quantum could be sufficient to break a much weaker bond between the donor and the electron, creating a free *photoelectron*. Or else it might help the electron to leave the atom of the lattice for the acceptor impurity atom, thus forming a hole. In both these cases it is said that there is an *impurity photoeffect.* The impurity concentration in a large majority of cases being much smaller than the concentration of the atoms in a semiconductor, the coefficient α corresponding to the impurity absorption is expected to be much smaller than in the case of the intrinsic absorption.

So, in the range of small values of λ (the large values being E_{ph}) the absorption is expected to be great. With λ increasing up to the values $\lambda_b = hc/E_g$, corresponding to the red boundary of the intrinsic absorption, the value α will diminish sharply and will remain small with a further increase of λ.

Have our expectations regarding the dependence $\alpha(\lambda)$ been answered? In general, yes, they have. But certain very important nuances are not considered.

Figure 37 shows $\alpha(\lambda)$ dependence for *p*-type silicon at room temperature. The dependences $\alpha(\lambda)$ of many other semiconductors have qualitatively the same shape, among them the *n*-type silicon, both *n*- and *p*-type germanium, gallium arsenide, and indium phosphide. Let us analyze this dependence.

Intrinsic Absorption. The energy of the electron-hole pair generation E_g in silicon is equal to 1.1 eV. So, with the wavelength $\lambda_l \approx 1.13$ μm the conditions $E_{ph} = E_g$ are satisfied. It is seen from Fig. 37 that in the region of intrinsic absorption, i.e. in the short wavelength region ($\lambda < \lambda_b$) the dependence $\alpha(\lambda)$ has been described by us correctly. With $\lambda = \lambda_b/2 \approx 0.5$ μm the absorption coefficient will make $\sim 10^4$ cm^{-1}. The wavelength increasing, the absorption coefficient will sharply decrease, and with $\lambda \approx 1.3$ μm the value α will be only ~ 0.36 cm^{-1}, even if the silicon is not very well purified ($N_a = 4 \cdot 10^{16}$ cm^{-3}).*

But our prediction of the variation $\alpha(\lambda)$ with the further increase of the wavelength λ proved to be wrong. With the increase of λ the absorption coefficient *increases* monotonously. For a given wavelength the value α

*In accordance with Eq. (48) and the dependence $\alpha(\lambda)$ shown in Fig. 37, the silicon plate with a thickness $d = 1$ mm will weaken the green light ($\lambda \approx 0.5$ μm; $\alpha \approx 10^4$ cm^{-1}) by a factor 10^{430}, the red light ($\lambda \approx 0.7$ μm, $\alpha \approx 2 \cdot 10^3$ cm^{-1}) by a factor 10^{90}, and the infrared light whose wavelength $\lambda \approx 1.2$ μm will penetrate through such a plate, practically without any absorption at all. The silicon plates (like the plates of germanium and gallium arsenide) are often used as simple and effective optical filters which cut off completely any short wavelength radiation.

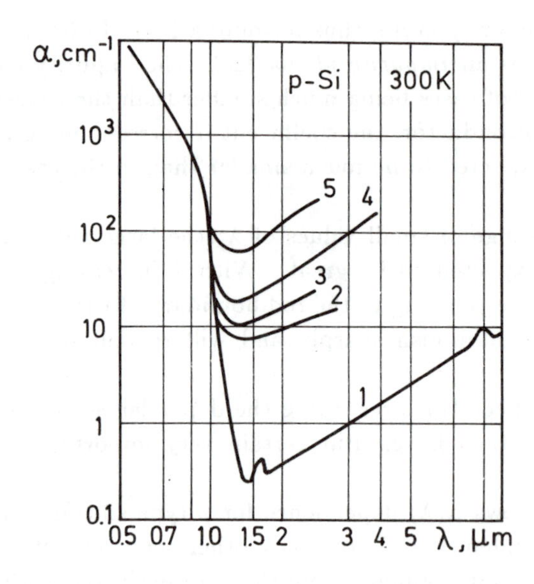

Fig. 37. The typical dependence of the absorption coefficient α on the wavelength λ at room temperature (p-type silicon is taken as an example). Different curves correspond to different levels of doping. Shallow acceptor impurity (boron) being introduced in the specimen: $1 - N_a = 4 \cdot 10^{16}$ cm^{-3}; $2 - 3 \cdot 10^{17}$ cm^{-3}; $3 - 8 \cdot 10^{17}$ cm^{-3}; $4 - 2.5 \cdot 10^{18}$ cm^{-3}; $5 - 7 \cdot 10^{19}$ cm^{-3}.

is proportional to the concentration of the doping impurity, and at the impurity concentration $\sim 10^{19}$ cm^{-3} even the minimal value α will make $\sim 10^2$ cm^{-1}.

What circumstance then have we not considered?

The Absorption by Free Carriers. The fact is that while doping the crystal with a shallow impurity, the temperature being high enough, all the impurity atoms get ionized on account of the thermal vibrations of the crystal lattice (recall the impurity saturation!). So on one hand, the impurity atoms being already ionized, the quanta of light cannot ionize them any longer. On the other hand, photons can be absorbed by the free electrons and holes in the crystal. The light absorption by the free carriers causes the monotonous increase of α with the growth of λ at $\lambda \gtrsim \lambda_b$.

This mechanism of absorption deserves a discussion though it has no relation to the intrinsic photoeffect: the light absorption by the free carriers *not being photoactive.* (The disappearance of the photon causing the appearance of neither an electron nor a hole. Hence, the crystal conductivity is not changed.)

First of all it should be mentioned that an *actually free carrier*, say an electron in the vacuum, in general cannot absorb a photon!

If the photon had been absorbed by the electron, the latter would have received the photon energy $E_{ph} = h\nu$ and, consequently, the velocity v, defined by the obvious equality $E_{ph} = m_0 v^2/2$ (where m_0 is the free electron mass in the vacuum). The electron moving with the velocity v has the momentum

$$p_0 = m_0 v = \sqrt{2E_{ph}m_0} = \sqrt{2hm_0\nu} \, . \tag{50}$$

From where does the electron get that momentum? Could it have received it from the photon it had absorbed? Let us explore this possibility.

The momentum of the photon p_{ph} is equal to

$$p_{ph} = m_{ph}c = (E_{ph}/c^2) \cdot c = h\nu/c = h/\lambda \, . \tag{51}$$

The ratio of the momentum of the electron which has absorbed the photon whose energy is $E_{ph} = h\nu$ to the momentum of such photon is equal to

$$p_0/p_{ph} = c\sqrt{2m_0/h\nu} \, . \tag{52}$$

By substituting into Eq. (52) the values of the velocity of light in vacuum c, of Planck's constant h and of the free electron mass m_0, this equation can be rewritten like this

$$p_0/p_{ph} \approx 1.6 \cdot 10^{10}/\sqrt{\nu} \, . \tag{53}$$

It is not difficult to calculate that the frequency of the electromagnetic wave, corresponding, say, to the green light ($\lambda = 0.5$ μm) is equal to $\nu = c/\lambda \cong 6 \cdot 10^{14}$ Hz. Thus, the electron which has absorbed a quantum of green light, according to Eq. (53), possesses a momentum 650 times greater than that of the absorbed photon. That means that the momentum was received not from the photon. So from where does the electron get it? It could not have been taken from the vacuum. Meanwhile when an electron absorbs a photon, the main laws of conservation – the laws of conservation of momentum and of energy – must be fulfilled just as in any other physical process. As we have seen, in the act of absorbing a photon by an electron in the vacuum, the laws of conservation of momentum and of energy cannot hold simultaneously. So? Such a process is absolutely impossible.

In a crystal things are quite different. The electron which has absorbed a quantum, can get the necessary momentum thanks to the thermal vibrations of the lattice (phonons) or to the interaction with the impurity centers. It is clear that the smaller the momentum the electron must borrow in order to absorb the photon, the greater is the probability that it will get it. The momentum decreases with the decrease of the photon energy (see Eq. (50)). That is why the absorption by free carriers grows with the increase of the wavelength of light, i.e. with the decrease of the photon energy (Fig. 37).

The Impurity Absorption. Curve 1 in Fig. 37 shows a small peak of the absorption coefficient in the region of the wavelength $\lambda \approx 9$ μm. The appearance of this spike is caused by the third mechanism of absorption which has already been mentioned by us – the impurity absorption. In the given case it is the absorption by oxygen, the impurity which is always present in silicon. The spike of the impurity absorption can be studied more thoroughly if the background, the absorption by the free carriers, is removed. In order to do it, it is necessary to cool the crystal to the temperature which is much lower than that of the impurity saturation. The electrons, undisturbed by the thermal vibrations of the lattice will be held tightly by the donor atoms. And the acceptors will not be able to take the electrons from the atoms of the lattice, forming holes.

There will not be any more free carriers in the crystal. So, there will not be any absorption by the free carriers either.

Under such conditions it is possible to investigate the absorption by the shallow donors and acceptors with a very small ionization energy. Figure 38 shows the spectral dependence of the absorption factor in the silicon cooled to the temperature of the liquid helium (≈ 4 K) in the region of the wavelength $\lambda \approx 10 \div 60$ μm (the photon energy being 0.02 – 0.12 eV). The maximum absorption in the photon energy region ≈ 0.04 eV corresponds to the ionization energy of the shallow acceptor, boron in silicon.

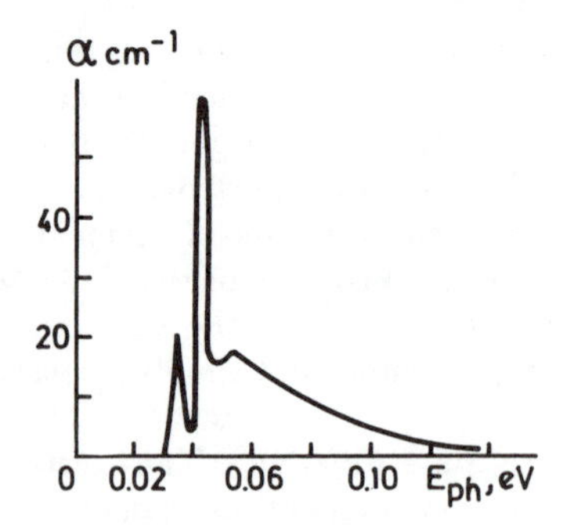

Fig. 38. Absorption coefficient α in silicon, with boron introduced in it, versus the photon energy E_{ph} ($T = 4$ K).

It is quite clear that when the photon energy E_{ph} is smaller than the impurity ionization energy ΔE there is no absorption whatsoever. But it is not clear why the absorption falls at $E_{ph} > \Delta E$.

The reason is in those very laws of conservation of energy and of momentum which prevent the free electron from absorbing a photon. With $E_{ph} > \Delta E$ the photon's excessive energy is to be given to the free electron. But that is impossible unless the electron receives simultaneously the corresponding momentum $p = [2(E_{ph} - \Delta E)/m^*]^{1/2}$ (compare it with Eq. (50) and recall that the motion of electrons in crystal is characterized by the effective mass m^*). The electron cannot receive the necessary momentum from the photon (see Eqs. (52) and (53)). It must be borrowed either from the thermal vibrations of the lattice (phonons) or from the impurity centers. The greater the necessary momentum, the more difficult it is to obtain it. That is why the impurity absorption falls with the increase of the photon excessive energy $E_{ph} - \Delta E$. Why then does the intrinsic absorption factor grow monotonously with the increase of the photon energy (Fig. 37)? The thing is that the disappearance of a photon energy during the intrinsic absorption brings about the appearance of not one, but two particles: an electron and a hole. The momentum conservation law holds in this case quite easily: if for example the effective masses of the electron and the hole are equal, and if after they appear, the electron and the hole travel with equal velocities in opposite directions, then the total momentum necessary to generate an electron-hole pair is just equal to zero.

The impurity absorption can often be *photoactive*. As a result of the absorption of a photon there appears a free electron (or a hole). The impurity photoactive absorption is widely used to produce photoconductors, sensitive in the infrared, where the photon energy is not large.

Now let us try and solve the following problem. Let us assume that all the semiconductors shown in Table 2 are at our disposal. Let us cut out a plate from each of them and look at a light source through it. What will we see?

The longest wavelength photon which can make man perceive color, is a "red" quantum with a wavelength $\lambda \approx 0.75$ μm. The energy of this quantum $E_{ph} \cong 1.65$ eV. It is seen from Table 1 that this value is much larger than the energy of the electron-hole pair generation E_g in InSb, Ge, Si, InP, and GaAs. Consequently, we will see nothing through the plates made of those semiconductors. Any quanta, corresponding to the visible light, will be absorbed even in a relatively thin layer of a semiconductor on account of a very strong intrinsic absorption.

The energy E_g in GaP (2.3 eV) is larger than the energy of the quanta of the red, orange and even yellow color ($\lambda_{\text{yellow}} \approx 0.55\ \mu$m, $E_{ph} \approx 2.25$ eV). But it is smaller than the energy of the quanta of the green color. Therefore we can expect the plate of GaP to absorb the quanta of the green color (to say nothing of the more energetic quanta of the blue and violet colors) and to transmit the red, orange and yellow colors. And really the GaP crystals look beautiful with their reddish orange shade. What then will be the color of the SiC crystal whose energy is $E_g = 3.2$ eV if we look through it? The energy E_g is larger than the energy of the violet quantum with the shortest wavelength ($\lambda \approx 0.4\ \mu$m, $E_{ph} \approx 3.1$ eV). Hence, such a crystal will transmit the whole spectrum of the visible colors and will be quite transparent. But such crystals are very rarely met. The majority of the SiC crystals look quite green against the light. They are colored (on account of the impurity absorption) by the nitrogen impurity which is practically always present in silicon carbide. (To obtain the right answer it is necessary to consider a lot of nuances, sometimes, most unexpected ones).

PHOTOCONDUCTIVITY

When electrons, holes or electron-hole pairs appear in a semiconductor under the action of light due to the intrinsic photoeffect, the conductivity increases.

The difference $\Delta\sigma$ between the conductivity of the illuminated semiconductor and the dark conductivity is called *photoconductivity*.

Under the action of light there appear carriers in a semiconductor which are excessive with respect to the equilibrium ones. Thanks to the process of recombination the excess carriers perish (recombine). If the generation and recombination rates are equal, then there is a stationary photoconductivity which does not depend on time.

If we switch off the light, the photoconductivity will diminish: the carrier concentration will decrease, tending to the state of equilibrium. The conductivity will then tend to the dark value, and the photoconductivity to zero. The characteristic time of the photoconductivity fall is determined by the lifetime of the excess carriers τ (see pages 75 and 99).

We can therefore watch the rapid change of the light intensity by means of a photoconductor only in the case when the lifetime of the excess carriers of the photoconductor material is short enough. The shorter the value of τ, the higher will be the speed of response of the photoconductor.

Should we conclude that the photoconductor must always be made of the materials whose values of τ are small? No. The smaller the value of τ, the lower is the sensitivity of the photoconductor.

The sensitivity of the photoconductor is determined in fact by the photoconductivity which appears in the semiconductor under the action of light. Let a certain number of excess carriers appear in a semiconductor under the action of light of a certain intensity. As soon as they appear they begin to perish due to recombination processes. The smaller the value of τ, the sooner they perish. It is clear that the smaller the value of τ, the sooner the carriers disappear, the smaller is the stationary photoconductivity $\Delta\sigma_{st}$ at a given light intensity. We may compare the process of achieving the stationary photoconductivity to that of filling the bath with water. A certain amount of water gets into the bath every second. If the drain hole is left open, the water begins immediately to leave the bath. The larger the drain hole, the quicker the water flows out of the bath, and the lower is the stationary level of water in the bath.

So photoconductors, designed to detect and measure very small light intensity changes, must be made of the materials whose τ values are rather large.

Photoconductivity Spectral Dependence

Figure 39 shows the photoconductivity-wavelength dependences of various semiconductor materials used to produce photoconductors. (Those dependences are called the photoconductivity spectral dependences.) It is seen that all the materials, without any exception, are characterized by a *selective spectral sensitivity*. That means that photoconductivity is created only by the action of light with the photon energy lying in a certain interval. At a certain wavelength of light, photosensitivity is maximal.

With the increase in the wavelength (the decrease of the photon energy) the photoconductivity falls. The reason is quite clear to us. In the first place, we know about the red boundary of the photoeffect. Secondly, we know of the sharp increase of light absorption near the red boundary of the photoeffect (Fig. 37). But looking at Fig. 39 we have an impression that there is not only the red but also the "violet boundary" of the intrinsic

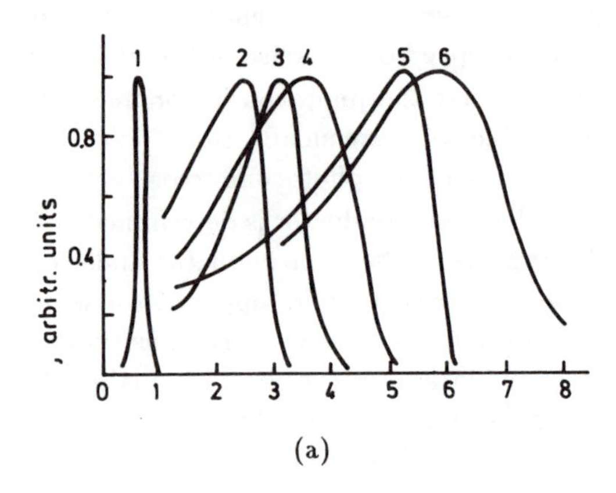

(a)

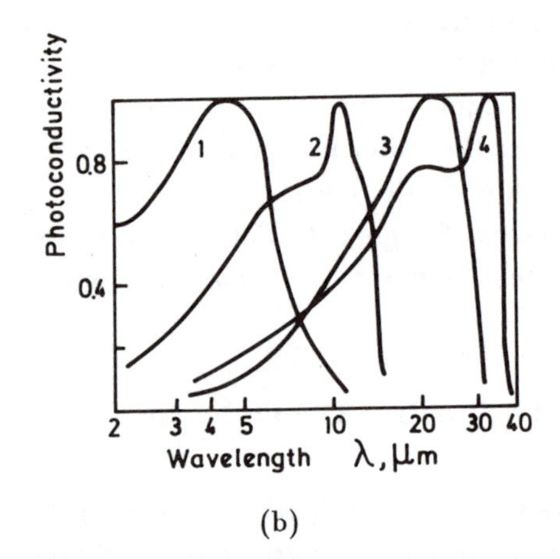

(b)

Fig. 39. Qualitative dependences of the spectral sensitivity of the photoconductivity for different semiconductor materials. (a) In the intrinsic conductivity region: 1 – CdS (300 K), 2 – PbS (300 K), 3 – PbS (77 K), 4 – PbSe (300 K), 5 – InSb (77 K), 6 – PbSe (77 K). (b) Impurity conductivity. The spectral sensitivities of photoconductors on the base of germanium with different impurities: 1 – Ge:Au (60 K); 2 – Ge:Hg (27 K); 3 – Copper impurity in germanium, Ge:Cu (15 K); Ge:Zn (4 K).

photoeffect. After photoconductivity reaches its maximum, it does not increase any longer with the decrease of the wavelength, but, on the contrary, it drops abruptly.

The thing is that with the decrease of the wavelength λ and the corresponding increase of the absorption coefficient α (look once more at Fig. 37), sooner or later, there will be a situation when the product of α and d (d is the thickness of the semiconductor plate) becomes greater than unity ($\alpha d > 1$). Then practically all the photons which have penetrated into the sample are absorbed there. (If you doubt it, study Eq. (49) again.) It is clear that if the condition $\alpha d > 1$ is satisfied and all the photons are absorbed in the sample, then the stationary photoconductivity $\Delta\sigma_{st}$ will not increase any longer with the increase of α. Then, with the decrease of λ, the conditions $\alpha d > 1$ being satisfied, we might expect the value $\Delta\sigma_{st}$ to remain constant. But as we know, with the further decrease of λ the photoconductivity drops. Why is it so?

It is because with $\alpha d \gg 1$, all the quanta of light are absorbed in a narrow layer close to the surface of the sample (Fig. 36, curve 3). The lifetime of the nonequilibrium carriers τ near the surface can be hundreds and thousands times shorter than the lifetime of electrons and holes in the volume of the semiconductor. There are certain traces of treatment on the crystal surface: grinding, chemical pickling and so on. Therefore the density of defects close to the surface is much greater than in the volume. And many defects act exactly like the deep levels, "the professional killers" of the nonequilibrium carriers (see Chapter 4). The greater the value of α (with $\alpha d \gg 1$) the closer to the surface are the electrons and holes born by light. Besides, when the electrons and the holes are born in a very shallow layer and their concentration is very great, they begin to collide with each other. Under such conditions, apart from their recombination via deep centers, the nonequilibrium carriers also disappear due to the immediate recombination of electrons and holes.

So, with $\alpha d > 1$ the number of carriers generated in the sample does not change with the growth of λ, while the recombination rate increases. That results in the decrease of the photoconductivity.

PHOTOCONDUCTORS

We have toiled a great deal, tilling the rocky soil of the intrinsic photoeffect; now the ripe fruit in the form of photoconductors is in our hands.

Figure 40 shows some types of photoconductors. They differ from each other in their shape, size, material and design. Nevertheless, having studied the physics of the photoeffect and photoconductivity, we know much about them. And what is not less important, we know what questions we are to answer to make things clear.

The base of any photoconductor is a semiconductor plate or film. If the photoconductor is designed to work in the intrinsic photoconductivity region, the maximum spectral sensitivity of the photoconductor will correspond to photon energies, approximately equal to the pair generation energy E_g in the given semiconductor (Fig. 39(a)).

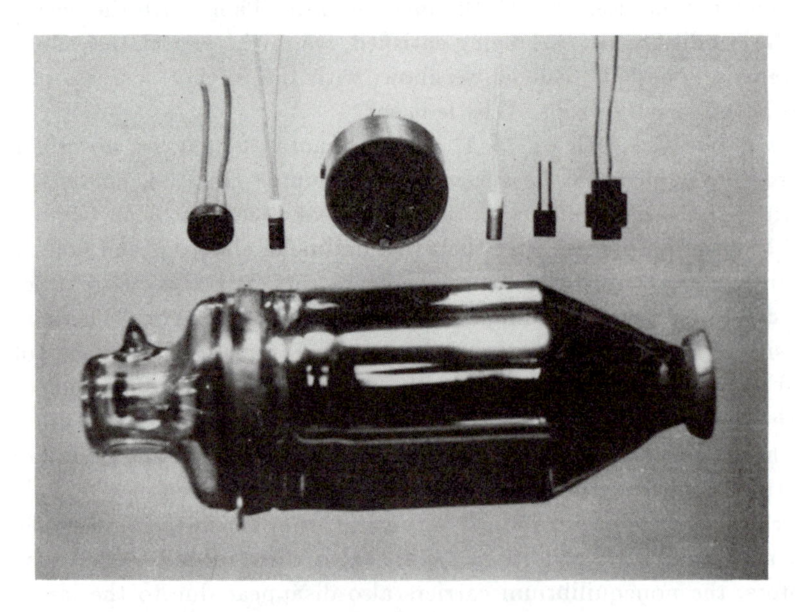

Fig. 40. Different types of photoconductors.

If the photoconductor is designed to work in the impurity photoconductivity region, the spectral sensitivity maximum will correspond to the photon energy approximately equal to the impurity ionization energy ΔE. The smaller the impurity ionization energy, the longer wavelength radiation will be registered by the photoconductor, and the lower the photoconductor operating temperature must be to avoid the impurity ionization due to the thermal vibrations of the lattice. It is necessary to cool the photoconductor designed to detect the long wavelength radiation.

The shorter the lifetime of the nonequilibrium carriers, the quicker will be the photoconductor response to the luminous flux change and the smaller the sensitivity, all other conditions being the same.

What Photoconductors Can Do

In the mid-seventies, one of the journals published a color photograph taken from the sputnik. It was not just a color slide but rather a "thermal photograph". A photoconductor, made of the semiconductor material (CdHgTe) with a negligible electron-hole pair generation energy E_g, was cooled to the temperature of liquid helium. The photoconductor registered with high precision the temperature of the terrestrial objects by the thermal infrared radiation emitted by them. The electron-optical system transformed the photoconductor signals into the color picture. The higher the temperature at a certain spot on the Earth's surface which the photoconductor was directed to, the stronger was the intensity of blue at the corresponding spot in the color "photograph".

The picture was taken in late autumn, when the ground was covered with snow. The greater part of the photograph was just pale-blue almost white, plains. On that pale blue background there were bright blue spots of settlements and small towns. A wide river, not yet covered with ice, wound its way throughout the picture. The river looked strange: up to a certain point it was of a quiet blue color; further on, downstream, the color was much darker, much more saturated. Then it gradually returned to its former quiet blue shade. The place where the river changed its color is marked with an arrow. There is an inscription beneath it: "At this point of the river the water is 0.5 degrees warmer. Only the waste of a very power-consuming plant can heat such a large river by half a degree. Most probably there is a secret plant there, producing concentrated plutonium."

Photoconductors, installed on sputniks, can detect the launching of foreign missiles. The launching is accompanied by an enormous flash of optical and thermal radiation. The analysis of the flash enables us to get an idea of the type of missile launched.

Photoconductors, capable of reacting to thermal radiation, are widely employed for peaceful uses: to measure the temperature of melted steel or pig iron in the metallurgical industry, of the incandescent mass in producing ceramics and cement, etc. Devices for measuring hot body temperature by the intensity and spectral composition of thermal radiation are called *pyrometers.* Pyrometers employing photoconductors are capable of measuring the temperatures about ten times lower than those measured by optical pyrometers.

Photoconductors made of semiconductors whose spectral characteristic of photoconductivity has its maximum in the visible-light spectrum (CdS, CdSe), are used in the devices measuring the levels of both natural and artificial illuminance.

Photoconductors are widely used in automatic security systems of premises. A beam of light, passing along the perimeter of the guarded territory, falls on the photoconductor. When the beam of light is intersected by some intruder the resistance of the photoconductor increases abruptly and then there is an impulse, which switches on the alarm. To guarantee the secrecy of the system an infrared light is used, as well as a modulation of the light beam; the intensity of the beam changes rapidly according to a certain modulation code. So no intruder can deceive the system illuminating the photoconductor with his own infrared source of light.

The change of the photoconductor resistance caused by the intersection of the illuminating beam is widely used in numerous counters of the produced articles at conveyors, in frequency meters controlling and regulating the rotation of engines, in protection devices and enclosures of machines and mechanisms and punch card reading devices in computers.

Photoconductors serve as sensitive elements in *nephelometers.* English for the Greek word "nephele" is cloud. Nephelometers determine the turbidity factor in liquids, suspensions and colloidal solutions. The operating principle of a nephelometer is very simple. The light flux from the calibrated light source having passed through the turbid medium, falls on the photoconductor. The higher the turbidity factor of the medium, the less light to fall on the photoconductor, hence the higher the photoconductor resistance. In nephelometers as well as in atmospheric pollution meters employed in mines and foundries, a circuit of the Witston bridge with two photoconductors is very often used. A beam of light reaches the first (comparative) photoconductor after having passed through the standard cell with a designed turbidity factor. If the turbidity of the environment is the same, then the light incident on the operating photoconductor has an intensity equal to that of the light incident on the comparative photoconductor. Then, both photoconductors having the same resistance, the output signal of the bridge is equal to zero. (See page 128). With the deviation of the environment transmittance from the standard, the bridge is unbalanced. The level of the output signal shows the degree of the environment turbidity. The universality and diversity of the abilities of a photoresistor are not inferior to those of a thermistor. They may even compete with each other. We know of the use of the thermistor in fire alarm systems. Though sometimes it is better to remember that "there is no smoke without fire" and not to wait for the temperature to rise higher causing the thermistor signalling system to work. The smoke filling the house, weakens the signals of the special electric lamp on the photoconductor. The smoke content reaching a certain level sets off a fire alarm signal.

Photoconductors are used to measure the amplitude and frequency of vibrations, the velocities and geometrical sizes of objects and their displacements: photoconductors being able to register displacements as small as a tenth of a micrometer.

Main Parameters of Photoconductors

Some of the main parameters of photoconductors are defined by the properties of the semiconductor material they are made of. We are practically acquainted with those parameters, such as the photoconductor dark resistance, the spectral sensitivity range and the response time.

The dark resistance R_d is defined by the dark conduction of the semiconductor material and its sizes. For various photoconductors it is within the range of $\sim 10^2$ to 10^8 ohms. Small values of R_d are characteristic of the photoconductors with a high speed of response. Large values of R_d are typical of relatively slow but sensitive photoconductors.

The range of spectral sensitivity, i.e. the range of wavelengths the photoconductor is sensitive to, including the wavelength corresponding to the maximum sensitivity, is defined mainly by the fundamental properties of the semiconductor material (Fig. 39), such as the energy of the electron-hole pair generation E_g for the photoconductor working in the intrinsic absorption region, and the impurity ionization energy ΔE for the photoconductor where the impurity absorption is used.

The photoconductor speed of response is characterized by the constants of the decay and rise time of the photocurrent τ. The values τ for photoconductors of various types lie in the range of $\sim 10^{-2}$ to 10^{-12} s. The fastest photoconductors with $\tau \sim 10^{-12}$ s are used to make physical experiments in which it is necessary to register the rapidly changing luminous fluxes.

It might seem easy to produce a material whose lifetime τ is short. One might think it would be enough just to introduce many deep impurities into it. But in fact it is not simple at all. Every impurity has its own limit of solubility in a given material. Introducing large amounts of impurity may spoil greatly some other properties of the semiconductor. It may decrease the mobility and may lead to the appearance of structural defects and so on. So, for the values τ to be very small, it is necessary to use specific technical methods, such as irradiating the semiconductor with protons or with other energetic particles.

There is another method, simple and fine, to create photoconductors with speedy response. It is based on the fact that the carriers, created in a semiconductor by light, are not allowed "to die their own death". A large voltage is applied to the photoconductor contacts, yet the distance L between them is very small ($L \leq 1$ μm). The field in the bulk of the photoconductor being very large, the velocity of the carriers moving in the device approaches its limit i.e. the saturation velocity (see Fig. 24). For the majority of semiconductors it makes $v_s \sim 10^7$ cm/s. So it takes the carriers created by light the time $t \sim \frac{L}{v_s} \sim 10^{-11}$ s to leave the bulk of the photoconductor.

The importance of other parameters is determined by the applications of the photoconductor. Though there is a great variety of circuits and devices in which photoconductors are used there are two main types which can be specified.

The first type comprises the photoconductors used in the circuits with a special panel lighting (counters, photorelays, angular accelerometers and displacement meters, security systems etc. For such photoconductors operating at large luminous fluxes, the decisive parameter is the resistance change factor K (with a certain exposure level). K is equal to the ratio of the dark resistance R_d to the resistance of the illuminated photoconductor. The typical value of K is in the range of $\sim 10^2$ to $\sim 10^3$.

The photoconductors of the second type must be capable of detecting the weak levels of the luminous fluxes. Their main parameter is, accordingly, that of the detective ability ("detectivity") and it is usually denoted by the letter $\mathcal{D}^*$.

$\mathcal{D}^*$ is inversely proportional to the minimum power the photoconductor is able to detect.

The parameter $\mathcal{D}^*$ is a complicated criterion in which a number of parameters of a photoconductor are considered. One of them is the intrinsic noise of a photoconductor, hindering the detection of very weak luminous fluxes. We will just compare the detectivity of the real and ideal photoconductors, the latter being absolutely noiseless.

One might think that such a photoconductor would detect any luminous flux, including the weakest one. But that supposition proves to be wrong. At the nonzero ambient temperature, the photoconductor is acted upon by the quanta of the background radiation. Even in the absence of any special source of radiation the background radiation generates free carriers in the photoconductor. These excess carriers change the conduction of the crystal. The background radiation fluctuates at random and these fluctuations prevent discovery of the weak radiation one wishes to detect.

The distribution of energy in the radiation spectrum is a function of the background temperature, i.e. of the ambient temperature. The spectrum of the radiation of bodies which corresponds to room temperature (300 K), is shown in Fig. 41. It is seen from the figure that the maximum radiation energy lies in the infrared ($\lambda \sim 10$ μm). There are practically no such photons in the spectrum whose wavelengths correspond to the visible region ($\lambda \sim 0.4 - 0.7$ μm). We are well aware of it. At night, when bodies do not shine in the reflected light of the sun, it is easy to bump one's head in the dark though the bodies around us "shine" most intensely with the infrared light.

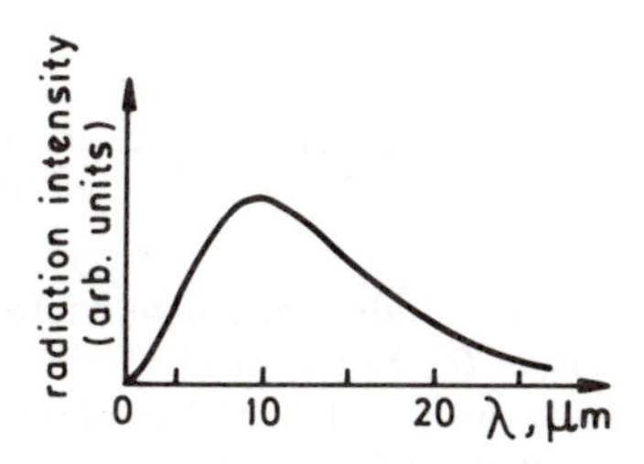

Fig. 41. Radiation intensity versus wavelength at 300 K.

Let us assume that we have a photoconductor at our disposal, made of a semiconductor with a large amount of energy of an electron-hole pair generation, say of gallium phosphide ($E_g = 2.3$ eV). The wavelength corresponding to the red boundary of the intrinsic photoeffect is equal to

$\lambda_b \cong 1.24/E_g \approx 0.54\ \mu$m. Photons with the energy $E_{ph} < E_g$ cannot generate electron-hole pairs in such a photoconductor. But in the background spectrum, with $T = 300$ K there are practically no quanta of light with the energy $E_{ph} > E_g$. Therefore the background radiation, corresponding to room temperature, will create negligible noises in such a photoconductor. Hence the ideal photoconductor made of the material whose E_g is very large will have a very high detectivity $\mathcal{D}^*$.

If the material for the photoconductor has been chosen with a smaller E_g (i.e. with a larger λ_b), then the larger part of the quanta emitted by the environment will cause the generation of electron-hole pairs in a semiconductor. The noise level, created by the background, will rise, while the detectivity will drop. With the decrease of E_g in a semiconductor (with the increase of λ_b) the detectivity $\mathcal{D}^*$ will diminish.

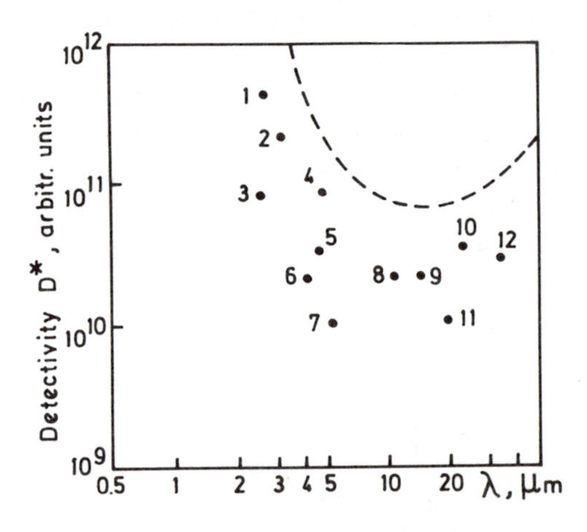

Fig. 42. Detectivity of the real and ideal photoconductors. The dashed curve indicates calculated dependences $\mathcal{D}^*(\lambda)$ for an ideal photoconductor. The dots indicate the maximum values of the detective ability of real devices. 1 – PbS (at the temperature of the working element $T = 200$ K); 2 – PbS ($T = 77$ K); 3 – PbS (300 K); 4 – InSb (77 K); 5 – PbSe (77 K); 6 – PbSe (195 K); 7 – Ge:Au (60 K); 8 – Ge:Hg (27 K); 9 – Ge:Cd (4.2 K); 10 – Ge:Cu (4.2 K); 11 – Si:Sb (4.2 K); 12 – Ge:Zn (4.2 K).

But if λ_b of a semiconductor becomes appreciably larger than the wavelength corresponding to the maximum heat background radiation ($\lambda \sim 10\ \mu$m, see Fig. 41), then practically all the photons, emitted by the environment, will generate free carriers in the semiconductor and will contribute to the background noise. No further increase of λ_b of the semiconductor will raise the noise level. Under these conditions the detectivity will stop falling with the increase of λ_b.

Figure 42 indicates the theoretically calculated dependence of the detectivity $\mathcal{D}^*$ of the ideal photoconductor on λ_b of the semiconducting material. The dots in the figure indicate the maximum values of $\mathcal{D}^*$ for the real photoconductors made of different semiconductor materials. As it is seen from the picture, the detectivity of the real devices, especially when cooled, are not very far from the theoretically limiting values in a very wide wavelength range.

Chapter 8. THE HALL EFFECT

> "The use of the magnet is so wide-spread nowadays that there is no necessity to speak at length about it".
>
> From the foreword by E. Wright to the book, "*On Magnet, Magnetic Bodies and on the Big Magnet – the Earth*" by W. Gilbert (circa 1600).

Figure 43 indicates a semiconductor specimen whose shape proves very convenient when studying the effect discovered in 1879 by the American physicist Edwin Hall. The specimen has the shape of a rectangular parallelepiped. There are ohmic contacts 1 and 2 at the ends of the specimen. The current I measured by the ammeter passes through them. Four more point contacts are located on the lateral surfaces of the specimen (3, 4, 5 and 6). That is all.

Let the current flow through the specimen. Using a voltmeter we will measure the voltage between contacts 3 and 5 (or 4 and 6). What do you think the voltmeter will show? It seems it will show nothing. And that is just so, our intuition has not deceived us this time. Indeed, suppose we have measured the voltage between contacts 1 and 3. What will it be equal to? It must be equal to the voltage drop between contacts 1 and 3 caused by

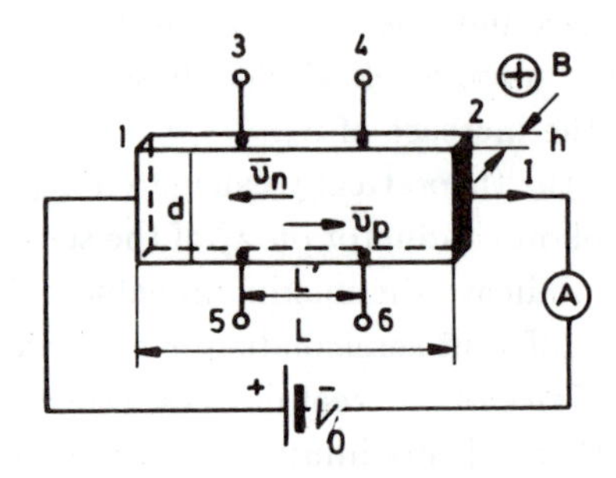

Fig. 43. In the magnetic field B, perpendicular to the plane, there is Hall emf between contacts 3 and 5 and between contacts 4 and 6.

the current flowing there. Now let us measure the voltage between contacts 1 and 5. It is clear that the reading of the voltmeter will be exactly the same, if contacts 3 and 5 are located exactly one over the other.

If the potential difference between contacts 1 and 3 V_{1-3} and between contacts 1 and 5 V_{1-5} is the same that means that the potential difference between contacts 3 and 5 (or 4 and 6) is equal to zero. These contacts are then said to be located on the line of the equal potential.

Now let us place the sample into the magnetic field directed perpendicularly to the figure and let us measure again the voltage V_{3-5} (or V_{4-6}). We will find the potential difference between these electrodes. The values V_{3-5} and V_{4-6} are proportional to the current I flowing through the sample and to the value of the induction of the magnetic field B. That effect was discovered by Hall and was named after him.

WHAT CAUSES THE APPEARANCE OF THE HALL EFFECT

Let us first discuss qualitatively the cause of the appearance of the Hall voltage between contacts 3 and 5 (or 4 and 6) with the application of the magnetic fields.

Let us assume that the sample under investigation is of n-type and the charged particles (electrons) drift in the direction opposite to that of the current. In Fig. 43 the current flows through the specimen from the left to the right. Thus, the electrons in the specimen move from right to left, from contact 2 to contact 1.

The moving carriers in the magnetic field are acted upon by the Lorentz force. The direction of the force can be found by the "rule of the left hand". When this rule is applied correctly, the turned thumb will show that the Lorentz force is directed upwards. Under the action of this force the electrons will flow up to the upper side of the specimen, charging it negatively, and will flow away from the lower side, charging it positively. Between the upper and lower sides of the specimen there appears the Hall voltage (Hall emf).

If the specimen is made of a p-type semiconductor and the current is conducted by a flow of holes, then, the direction of the current and of the field being the same, the Lorentz force will still be directed upward. But the motion of the holes towards the upper side of the specimen corresponds to the positive potential of the upper side with respect to the lower side. The Hall voltage (Hall emf) will have a sign opposite to that of Hall voltage in the electronic sample.

By the sign of the Hall voltage one can judge the conductivity type of the sample.

HALL FIELD

The Hall voltage which has appeared under the action of the Lorentz force prevents any further flow of electrons to the upper side of the specimen (at the chosen direction of the current and the magnetic field (Fig. 43)). At the stationary condition, which has been established, the Lorentz force acting on the carriers on the part of the magnetic field $f_L = q\bar{v}B$ will be counterbalanced by the force $f_H = qF_H$ acting on the carriers on the part of the Hall field F_H. The carriers will stop flowing up to the upper side of the sample as well as flowing away from its lower side and will continue moving along parallel to the long sides of the sample the way they did before the magnetic field was applied (Fig. 43). Thus the Hall field F_H will be determined from the equation

$$q \cdot \bar{v} \cdot B = qF_H \,. \tag{54}$$

Here $\bar{v} = \mu F_0$ is the drift velocity of the carriers in the field $F_0 = V_0/L$; V_0 is the bias voltage, L is the length of the specimen (Fig. 43).

Hence

$$F_H = \mu B \cdot F_0 \,. \tag{55}$$

At the given value of the bias field F_0 created by the external bias voltage and the given value of the induction of the magnetic field B the value of the Hall field F_H is proportional to the mobility of carriers μ.

Pay attention to the unusual situation which is illustrated in Fig. 44. The total intensity of the electric field in the sample $\mathbf{F}$ is equal to the vector sum of the field $\vec{F}_0$ and the Hall field $\vec{F}_H$. But the carriers move parallel to the field $\vec{F}_0$ and not to the total field $\mathbf{F}$. Why?

The force acting upon the electron on the part of the Hall field is counterbalanced by the Lorentz force equal to it in magnitude and opposite in direction. The angle Θ between the field F_0 and the total field is called the Hall angle. From Eqs. (54), (55) it is clear that $\tan\Theta = \mu B$.

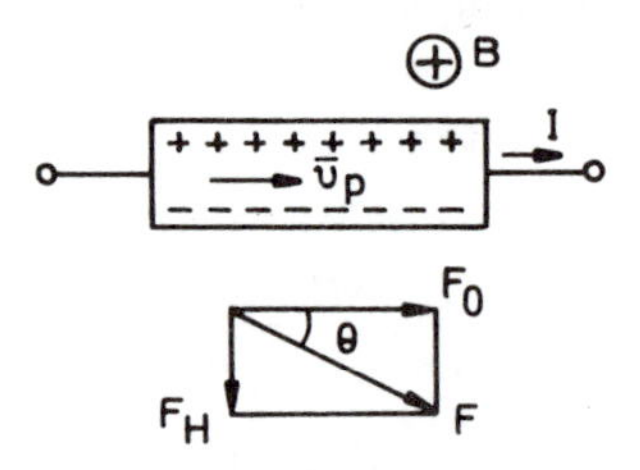

Fig. 44. The total electric field F is equal to the vector sum of the bias field F_0 and Hall-field $F_H \cdot tg\Theta = \mu B$. The force, acting on the carrier on the part of the Hall field F_H is balanced by the Lorentz force. (The sign of the Hall field in the figure corresponds to the hole conductivity of the specimen.)

WHAT IS MEASURED EXPERIMENTALLY

A researcher who studies the Hall effect in a semiconductor, measures the current I flowing through the sample, the induction of the magnetic field B, and the potential differences between electrodes 3, 4, 5 and 6 (Fig. 43). His aim (as well as ours) is to calculate the carrier density and mobility in a semiconductor, making use of the values which have been measured. Knowing the physical sense of the Hall effect, we can easily make all the necessary calculations.

Let us measure the conductivity of the sample σ without switching on the magnetic field. First we have to measure the potential difference between electrodes 3 and 4 ($V_{3\text{–}4}$) at the given current I. After which the conductivity is found from the equation

$$\sigma = \frac{I \cdot L'}{V_{3\text{–}4} \cdot h \cdot d} . \tag{56}$$

The meanings of all the symbols are clear from Fig. 43: L' is the distance between the contacts, d is the width, h is the thickness of the sample. But how has this equation been obtained? Where has it come from? The answer is quite simple. The specimen resistance between contacts 3 and 4 is equal to $R_{3\text{–}4} = \rho \frac{L'}{S} = \frac{L'}{\sigma \cdot h \cdot d}$. And the resistance $R_{3\text{–}4}$ is equal to $R_{3\text{–}4} = V_{3\text{–}4}/I$. Hence Eq. (56) is derived.

For contacts 5–6 analogous measurements and calculations can also be made. This enables us to check the uniformity of the specimen. If the measurements on contacts 3–4 and 5–6 give different values for σ, that means that the conductivity in different parts of the specimen is different.

Locating on the lateral sides not one, but several pairs of electrodes, we can investigate the distribution of non-uniformities in the specimen, and sometimes that enables us to help the technologists in obtaining uniform specimens.

Now let us switch on the magnetic field and measure the magnetic induction B and the Hall voltage V_H between contacts 3 and 5 ($V_{3\text{–}5}$), the current I in the specimen being the same. Now we just have to calculate. The Hall voltage V_H is equal, of course, to the Hall field F_H, multiplied by the width of the specimen d.

$$V_{3\text{–}5} = V_H = F_H \cdot d = \mu B F_0 \cdot d . \tag{57}$$

The value of the field F_0 is known. It is just equal to the potential difference between contacts 3 and 4, divided by the distance between them: $F_0 = V_{3\text{–}4}/L'$. Using Eq. (56), we obtain

$$F_0 = V_{3\text{–}4}/L' = I/h \cdot d \cdot \sigma . \tag{58}$$

Substituting (58) into (57), we get

$$V_{3\text{–}5} = IB\mu/h\sigma . \tag{59}$$

Recalling that $\sigma = qn_0\mu$ (see page 82), we have

$$V_{3\text{-}5} = IB/qn_0h \ .^* \qquad (60)$$

The values of the conductivity σ and carrier density n_0 being known, the mobility value μ will be defined as $\mu = \sigma/qn_0$.

We have already spoken about the importance of being able to determine the values of mobility and concentration. Using the Hall effect is the most common way of measuring those values in semiconductors.

APPLICATIONS OF HALL EFFECT

One of the most important applications of the Hall effect is the investigation of the properties of semiconductors, metals and some dielectrics. The Hall effect is of help when investigating the dependences of the concentration and mobility of carriers in different materials on temperature, pressure and on the type and concentration of different impurities. The dependences shown in Fig. 22 have been obtained by means of investigating the Hall effect.

Modern Hall installations for investigating the properties of semiconductors are rather complicated technical equipment. The temperature in the chamber where the specimen is placed can change automatically from a few degrees to several hundred Kelvins. To create a strong magnetic field, superconducting solenoids are used. They are cooled by liquid helium and allowed to obtain the magnetic induction of 80.000 $\div$ 100.000 Gs. To be able to measure small values of mobility ($\lesssim 10^{-6}$ m^2/(V·s.)), it is necessary to use complicated radiotechnical circuits which allow us to measure very small Hall voltages on the background of strong noises. The results of the measurements are transferred to a computer where the data are processed.

Such complicated installations are not numerous. But there are thousands of simpler Hall installations. The Hall effect is used in numerous

---

*It may seem that there is some contradiction between Eqs. (57) and (60). In one case the Hall voltage depends only on the mobility carriers, in the other, on their concentration. But there is no contradiction here whatsoever. In fact the smaller the density n_0 and the mobility μ, the greater must be the bias field F_0 applied to the specimen (see Eq. (58)) to let the given current I pass through it.

laboratories to control the quality of the semiconductor materials and devices. Besides, even rather simple installations for measuring the Hall effect make it possible to investigate physical phenomena.

The Hall effect has a lot of different technical applications. We will discuss only the most essential and interesting of them.

Measuring the Magnetic Field

Modern physics and techniques use magnetic fields whose induction is in the range of $\sim 10^{-3}$ to 10^7 Gs. It is necessary to be able to measure the values of the magnetic field at the temperatures from a thousandth of a degree to thousands of degrees, in the frequency range from a constant field to $\sim 10^{10} - 10^{11}$ Hz, corresponding to the microwave range.

Hall Sensors – that is the name of semiconductor devices whose action is based on the Hall effect. Hall sensors are widely used to measure the magnetic induction. A very large range of magnetic fields to be measured as well as the necessity of wide temperature range measurements require the use of various semiconductor materials to produce the Hall sensors. The following materials are commonly used: Ge, Si, InSb, InAs, HgTe, CdTe and GaAs. To measure weak magnetic fields at low temperatures it is necessary that Hall sensors should be made of materials with high mobility and the small energy of electron-hole generation E_g. For the measurements in high temperature regions, materials with large E_g are used.

The principle of measuring the magnetic field by the Hall effect is very simple: the Hall voltage in the sample with a given current flowing through the sensor is directly proportional to the magnitude of the magnetic induction (see Eq. (60)). A high degree of linearity (the precision with which the output signal of the sensor is proportional to the quantity which is being measured) is a very important advantage of the Hall sensors over any other sensors measuring the magnetic induction.

Measuring the Current

At first sight using the Hall sensor to measure the current may seem absolutely unnecessary, or at least, exotic. Indeed, there are tens, or even hundreds of devices, appliances and circuits to measure the current. In the first place it is certainly, ammeters of various systems, designed for the currents from picoampere (10^{-12} A) to kiloampere (10^3 A). Then there are measuring resistors – small resistors, connected in series to the circuit whose current is to be measured. The potential drop across the measuring resistor is proportional to the current flowing through it.

In modern engineering it is necessary to measure the direct current whose value is hundreds of thousands of amperes. Such currents are flowing in industrial electrolytic installations. Some pulse currents used in physical experiments are tens of millions of amperes. Besides, they have a very complicated time dependence which is also to be studied. No other additional loads, such as shunts or ammeters can be connected to the circuit in which a powerful current is flowing. Apart from consuming the power of thousands of kilowatts, such additional loads could distort the electrical parameters of the circuit. In some cases service conditions do not allow breaking the circuit and connecting in series any measuring resistor.

So what can be done?

The current can be measured by the intensity of the magnetic field created by it! And whenever it is necessary to measure the magnetic field – the Hall sensors, as we have just seen, are most convenient and sometimes even indispensable.

The time of establishing the Hall voltage is usually $\sim 10^{-11} - 10^{-12}$ s. This enables us to investigate even very short pulses. Due to a high linearity of the Hall sensors it is possible to measure directly the amount of the current and its time dependence. Due to the small size of the devices it is possible in case of necessity to measure the distribution of the current in space (for instance in electrolytic baths).

Measuring the Microwave Power

Measuring the microwave power by means of the Hall sensors presents an example of a fine and simple solution of a very complicated technical problem. The Hall sensors for measuring the microwave power need neither a magnet, nor a battery to create the bias field. The electromagnetic wave whose power is to be measured serves as the source of both: the bias and the magnetic fields.

Figure 45 shows the electromagnetic wave propagating in the direction **k**. The vector of the electric field **F** oscillates in the vertical plane, while the vector of the induction of the magnetic field **B** – in the horizontal plane. The Hall emf is created on the lateral sides of the Hall sensor placed in the

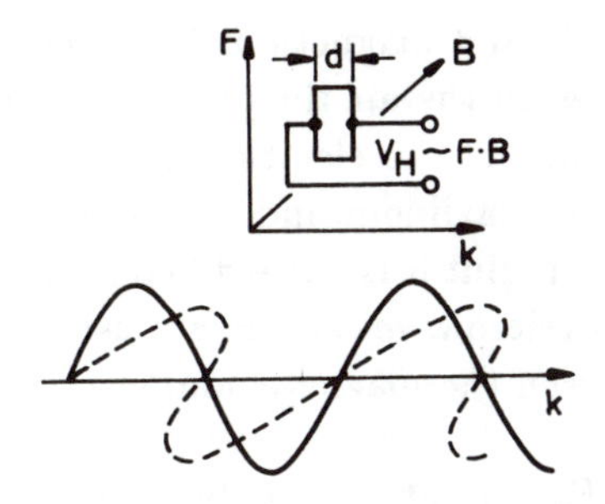

Fig. 45. The vector of the electric field **F** in the electromagnetic wave, propagating in the direction of the vector **k**, oscillates in the vertical plane (solid curve), and the vector of the magnetic induction **B** oscillates in the horizontal plane (dashed curve). There is Hall emf U_H on the lateral surface of the specimen placed in the field of the electromagnetic wave (the insertion).

field of the electromagnetic wave

$$V_H = \mu d \mathbf{F} \cdot \mathbf{B} \ . \tag{62}$$

In Eq. (62) the values of **F** and **B** are time dependent:

$$F = F_m \sin \omega t$$
$$B = B_m \sin \omega t \ .$$

The Hall emf V_H changes in time with the frequency twice as large as that of the electric or magnetic field

$$V_H \sim F_m \cdot B_m \sin^2 \omega t = \frac{F_m B_m}{2} (1 - \cos 2\omega t) \ . \tag{63}$$

In comparison with other devices for measuring the microwave power, the Hall sensors have two great advantages. The Hall emf indicates the direction of the propagation of the wave. As soon as the direction of the propagation of the wave changes to the opposite, the sign of the Hall voltage changes too. The direction of the vector **F** being the same, the change of the direction of the electromagnetic wave causes the vector **B** to change to the opposite direction. The second advantage of the Hall sensors over thermistors, bolometers and some other types of devices measuring the microwave power, lies in the fact that the Hall voltage is sensitive to the phase difference between the electric and magnetic field in the electromagnetic wave. When a wave propagates in the air the phase difference is always equal to zero. The changes in time of the electric and magnetic field are *cophased* i.e. the fields reach their maximum and pass through zero simultaneously (as it is shown in Fig. 45). But it is quite different in waveguides and in resonators where the Hall microwave sensors are usually used. There appears a phase difference between the magnetic and electric field

$$F = F_m \cdot \sin \omega t\,; \quad B = B_m \cdot \sin(\omega t + \varphi)$$

analogous to the appearance of a phase difference between the current and the voltage when the load contains the reactive elements – capacitance or inductance. The mean value of the Hall voltage depends on the phase difference between the magnetic and electric fields. It may be shown that the value of the mean Hall voltage is proportional to $\cos\varphi$, where φ is the phase difference between the magnetic and electric fields. The power absorbed in the load is also proportional to $\cos\varphi$. Therefore the Hall microwave power sensors allow maintaining the immediate control over the power absorbed by the load.

Using semiconductors with a high mobility μ we can design rather sensitive microwave Hall sensors. With the help of the antimonide-indium Hall sensors we can achieve a sensitivity corresponding to 0.1 mkV of the Hall voltage per 1 mW of the microwave power absorbed in the load.

Hall Sensors in Automatic Regulation Systems

The most popular and most important application of the Hall sensor is that in the systems of the automatic regulation.

Everyone knows how unpleasant it is if there are acoustic distortions in record players. It often happens if the motor of a tape recorder or of a record player is incapable of keeping the cassette or disk velocity constant. The reason for it may be the fall of the voltage across the source (if the voltage falls lower than the allowed value), or if the film is jammed, or if there is too much dust in the rotating parts of the device. Still more troublesome is the irregular advance of the tape in the video tape recorder. Yet the most aggravating phenomena are the consequences of the nonuniform motion of the computer disk reader. It is possible, of course, to find a temporary way out by trying to stabilize the power sources, by protecting the device from dust, by treating it most carefully etc. But with modern requirements for the regulation of the motor rotation in the most precise and complex machines, these precautions will not help much. The only radical way out in such cases lies in the systems of automatic regulation.

The rotating motor must "inform" a certain control unit, say, a microprocessor of its rotation velocity. This control unit will react accordingly, either by accelerating or by retarding the rotation of the motor.

The Hall sensors, miniature, simple and reliable, and with a great response speed, proved to be quite indispensable here. They are placed around the revolving motor shaft, on which tiny magnets are fixed. Each time the magnet passes by the Hall sensor it induces the Hall voltage pulses. The pulses arrive at the control unit (the microprocessor). If they arrive with the wanted speed, everything is O.K. If the frequency is lower than necessary, then the rotor slows down its spin, after which the control unit immediately increases the source voltage and restores the necessary rotation velocity.

The Hall sensors also regulate with great precision and reliability the ignition in automobile cylinders. Besides, the Hall sensors are often used in pairs with miniature magnets which are acting as "position pickups". For instance to automatically check if the door of the car is closed properly. Provided the magnet, hidden in the sheath of the door, approaches the Hall sensors incorporated in the frame, there appears the Hall voltage which "lets the microprocessor know" that the door is closed. Otherwise there appears a special signal: "Shut the door, please!"

Most probably, you yourself, without being aware of it, use quite a few Hall sensors in your everyday life. That accounts for the annual produce of millions of Hall sensors made of silicon or of gallium arsenide.

Chapter 9. GUNN EFFECT

About thirty years ago, in the early sixties, the American physicist John Gunn decided to investigate the current-voltage characteristics of semiconductors, comparatively new at that time, gallium arsenide (GaAs) and indium phosphide (InP), in very strong electric fields. He discovered a wonderful effect which was later named after him.

Gunn's results are shown in Fig. 46. Voltage pulses were applied to the samples (pieces of GaAs or InP with two ohmic contacts). If the pulse amplitude V was smaller than a certain threshold value V_t – different for different samples – then everything was clear. The current in the sample was growing while the voltage was growing, it was constant when the voltage was constant and it dropped when the voltage dropped. This case is shown in Fig. 46(a), (b) by a dashed line. By changing the voltage pulse amplitude and measuring the amplitude of the current corresponding to it, it was possible to obtain a current-voltage characteristic of the sample. It usually corresponded with Ohm's law.

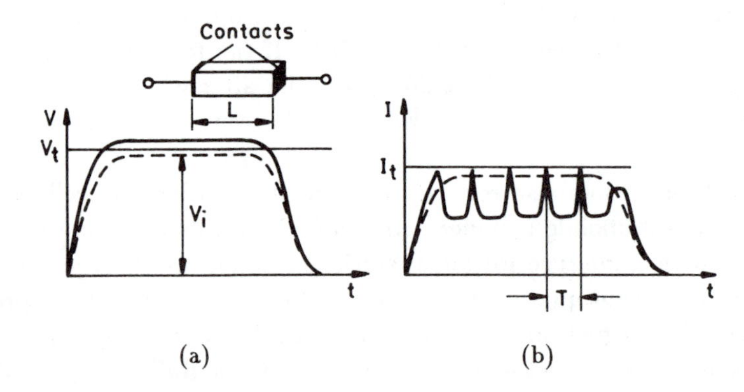

Fig. 46. Time dependences in Gunn's experiments: (a) The voltage across the specimen versus time. (b) The current versus time. The broken line indicates the case when the voltage pulse amplitude V is less than the critical value V_t. The solid line indicates the case when $V > V_t$. The shape of the specimen is shown in the insert.

But as soon as the voltage exceeded the value V_t (the index "t" coming from the word "threshold"), the current would begin to oscillate with the voltage being constant. This case is indicated in Fig. 46 by solid lines. Using the samples whose lengths were different, Gunn established that the threshold voltage was proportional to the length of the sample $V_t = F_t \cdot L$, where F_t was approximately the same for all the samples. For GaAs the value F_t was approximately $3.2 \cdot 10^5$ V/m.

The period of oscillation T was also proportional to the length of the sample L. It was approximately equal to $T \approx 2L/v_t$ where v_t is the velocity of electrons in the threshold field. Since Ohm's law is valid, roughly speaking, up to the threshold field F_t, the threshold velocity is $v_t \approx \mu F_t \approx 2 \cdot 10^5$ m/s (the mobility of the electrons in the GaAs samples, which were at Gunn's disposal, was at room temperature about $0.5 - 0.7$ m^2/(V$\cdot$ s)).

In 1963 Gunn described the results of his observations in a short paper which attracted the attention of dozens of physicists and engineers all over the world. Physicists were interested in the explanation of those phenomena, while engineers realized at once that the discovery of the Gunn effect meant the appearance of a new semiconductor generator, capable of working in a microwave band.

Indeed, with the length $L \approx 2 \div 10$ μm, the oscillation frequency, f, generated by a diode is

$$f = \frac{1}{T} \cong \frac{v_t}{2L} \sim (1 \div 5) \cdot 10^{10} \text{ Hz} . \tag{64}$$

This frequency range is of great relevance to radioelectronics, especially radiolocation and to the numerous communication systems both civil and military.

Though dozens of highly qualified specialists racked their brains over the nature of the phenomenon discovered by Gunn, a whole year had passed until it was explained.

During that year Gunn made a skillful investigation of the effect discovered by him. He established that when the voltage on the sample exceeded the threshold value V_t, the field distribution in the sample was changed completely. At $V < V_t$ the current in the sample is a flow of electrons from the negative contact (the cathode) to the positive contact (the anode). The electric field is distributed along the sample uniformly (Fig. 47, the dashed line). As soon as the voltage on the specimen exceeds the threshold value V_t, there appears at the cathode a narrow layer of a very strong field, a "domain".* Though the layer is not wide, the field is so strong there, that a very large part of the voltage, applied to the sample, drops on the domain (Fig. 47). The domain moves from the cathode to the anode with a velocity approximately equal to $v_t/2$. On reaching the anode contact, the domain

---

*In physics the word "domain" is used to denote such regions whose properties differ greatly from those of the environment. There are ferromagnetic and ferroelectric domains, domains of strong or weak electric fields... In Fig. 47 a strong electric field domain is shown.

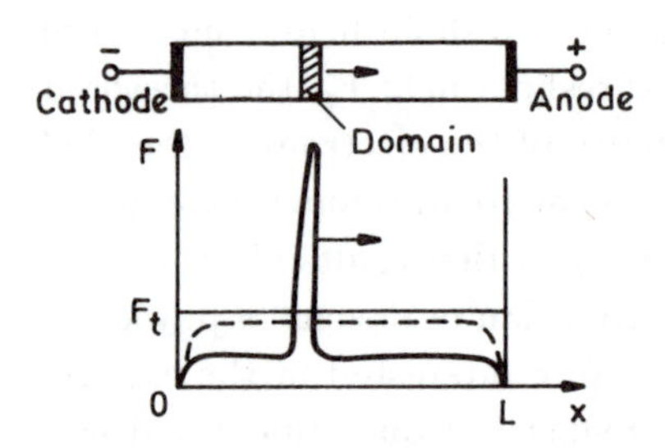

Fig. 47. The distribution of the electric field in the specimen. The broken line indicates the case when $V < V_t$. The solid line indicates the case when $V > V_t$. The arrow shows the direction of the domain motion: from the cathode to the anode.

decays. Immediately after that, a new domain is formed at the cathode and everything is repeated all over again.

It is easy to see the connection of the domain's periodic motion through the sample with the current oscillations which had been observed by Gunn in his experiments.

Let us assume that a voltage, somewhat smaller than the threshold value V_t, is applied to the sample (Fig. 46(a), the dashed line). Then the current, a bit weaker than I_t, flows through the specimen, and the field, less than the threshold value $F_t = V_t/L$ (Fig. 47, the dashed line), is distributed along the specimen uniformly. It has already been mentioned that while the voltage V is less than V_t, Ohm's law is roughly fulfilled. That means that the velocity of the electrons in the specimen is a bit less than the value $v_t \cong \mu F_t$.

Now let us increase the voltage applied to the specimen to such an extent that it would be equal to V_t. There will appear a domain in the specimen, and a large part of the voltage on the domain will drop. But that means that the voltage on the other parts of the specimen, where there is no domain, will now be less. The field outside the domain has decreased (cf. the dashed and solid curves in Fig. 47). The velocity of the electrons which was proportional to the field has dropped too, so has the current flowing through the specimen.

The current does not change while the domain moves along the specimen. On reaching the anode, the domain begins to decay, and then the voltage drop across the domain become less. Meanwhile the field in the other part of the specimen increases, and so does the current. But as soon as the field in the vicinity of the cathode of the specimen becomes equal to the threshold value F_t, a new domain is being formed in the sample and the current is decreasing. This process is repeated periodically (Fig. 46(b), the solid curve), until the voltage applied to the current stops exceeding the threshold value V_t.

AMAZING DEPENDENCE OF THE VELOCITY OF ELECTRONS UPON THE ELECTRIC FIELD

The wonderful experiments made by Gunn enabled us to see the processes which the current oscillations in the sample result in. But they could neither answer the question why there was a domain in the sample, nor explain the physical sense of the appearance of that phenomenon.

Those questions were answered in 1964 by the American physicist Herbert Kroemer. He drew the researchers' attention to the fact that not long before Gunn had made his discovery it had been proved by means of theoretical calculations that in gallium arsenide, indium phosphide and in some other semiconductors the dependence of the drift velocity of electrons on the electric field must look exactly the way it was indicated in Fig. 48. Within a certain range of the fields F the drift velocity of electrons $\bar{v}$ *decreases with the growth of the field.* As we shall see later, this kind of $v(F)$ dependence offers every explanation for the phenomena observed by Gunn. But what makes this characteristic $v(F)$ possible?

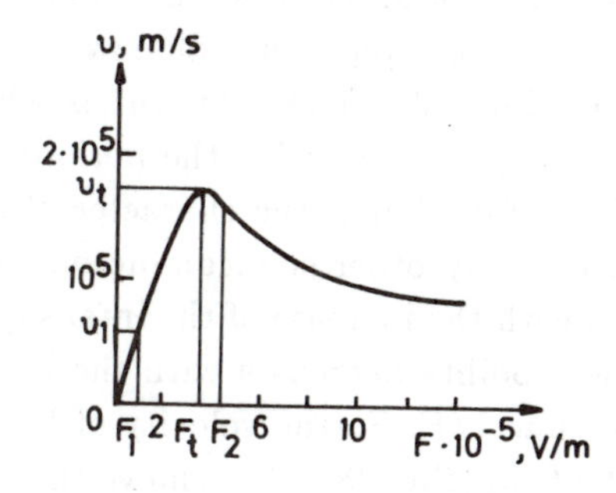

Fig. 48. Electron drift velocity ν versus the intensity of the electric field F in GaAs. Similar dependences are typical for many semiconductors (InP, GaInSb, InGaAsP etc).

The Decrease of Mobility

When discussing "hot electrons" in Chapter 4 it was mentioned that the main parameters of the hot electrons, determining the carrier mobility – the scattering time τ_0 and the effective mass m^* – are functions of the electric field. The character of those dependences is determined by such factors as the type of the semiconductor, the temperature of the crystal lattice, the type of impurities and also the carrier energy depending upon the electric field.

In germanium and silicon, as we recall (Fig. 24), the drift velocity of electrons is saturated with the growth of the field. In GaAs, as it is seen from Fig. 48, the velocity even falls with the growth of the electric field. To make it possible, the decrease of the carrier mobility with the increase of the field must be still faster than in Ge and Si. This is realized in GaAs, InP and in some other semiconductors. In those materials the growth of the electric field causes not only a decrease of the scattering time of hot electrons, but also an increase of the electron effective mass m^*. We have already met with the carrier effective mass change. In strain gauges the carrier effective mass is a function of pressure due to the change in the mutual arrangement of the lattice atoms. This altered arrangement changes in the fields created by the lattice ions and the atomic valent electrons.

In a strong external electric field where the drift velocity falls with the growth of the field (in GaAs) the mutual arrangement of the lattice atoms does not change. What changes is the character of the interaction between the electric fields created by the lattice atoms and hot carriers. Let us recall that the effective mass of an electron and a hole m^* is not equal to the free electron mass in vacuum m_0. The value m^* can differ from that of m_0 dozens of times (see Table 3). The value m^* is wholly defined by the interaction of the electron (the hole) with the periodic field of the crystal lattice. The value m^* is a function of the character of their interaction.

In GaAs, InP and in many other semiconductors the effective mass of the hot electrons grows with the increase of the intensity of the electric field F so abruptly that the mobility decreases with the increase of the electric field faster than by the law $1/F$. So the velocity of the electrons decreases with the increase of the field (Fig. 48). To achieve this effect it is necessary to heat the electrons to a great extent by the electric field. The higher the carrier mobility, the sooner the electrons acquire the energy in the electric field, and the smaller is the value of the threshold field F_t. So, the value

F_t for GaAs is $\sim$ 3 kV/cm, for InP it is $\sim$ 10 kV/cm, and for InSb it is only 0.5 kV/cm. (Compare these values with the mobility values given in Table 2).

EVERYTHING IS RATHER SIMPLE IN FACT

What is the connection of the phenomena observed by Gunn with the dependence $v(F)$ shown in Fig. 48?

To answer this question let us first consider a very simple problem, which on the face of it seems to have no relation whatsoever either to the Gunn effect or to the specific properties of hot electrons.

The Fluctuations of the Field on the Ohmic Part of the Dependence $v(F)$ Disappear

Let the voltage $V_1 < V_t$ be applied to a sample (Fig. 48). The field is distributed uniformly and it is $F_1 = V_1/L$. The drift velocity of electrons $\bar{v}_1$ is equal to μF_1 and the current is a flow of electrons from the cathode to the anode with a velocity $\bar{v}_1$.

Now, let us assume that there is a small region in the specimen where the field is a bit larger than F_1. Due to the chaotic character of the motion of electrons in the crystal, there constantly appear slight deviations of the intensity of the field from the average value F_1 (the fluctuations of the field). Our task is to observe such a fluctuation from the moment it appears.

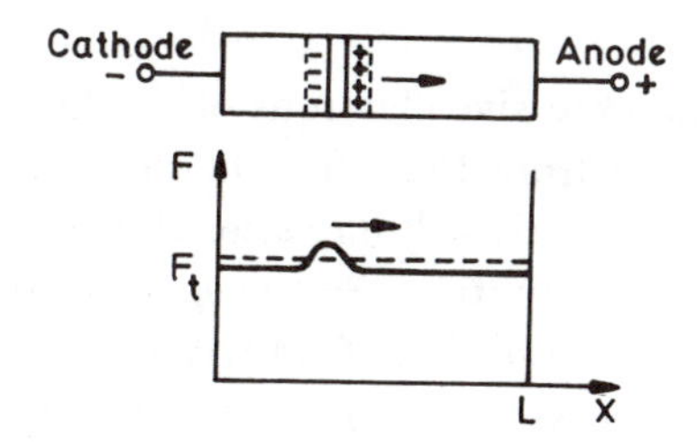

Fig. 49. The distribution of the electric field in the specimen without fluctuation (dashed line) and with fluctuation (solid line). A diagram of the space charge on the borders of the fluctuation is given in the upper part of the figure.

If in some part of the specimen the field has become somewhat larger than in the rest of the specimen, that means that electric charges have been stored in the borders of that region: negative charge on the cathode side and positive charge on the anode side (Fig. 49). The field of that excess space charge, added to the field of the external source, has somewhat strengthened the field within the considered region.

Since we are considering the case when the field F_1 is weaker than the field F_t, the velocity of the electrons therefore is proportional to the field (Fig. 48). Consequently, the electrons within the fluctuation move faster than the electrons in other parts of the specimen outside the fluctuation. That means that the fluctuation will "run" from the excessive electrons on the left towards the region on the right where there is a deficiency of electrons. That will result in the disappearance of the fluctuation.

The Fluctuations Increase and Transform into a Domain in the Case When the Electron Velocity Decreases with the Increase of the Field

Let us return to the Gunn effect and see what will happen to the fluctuation of the field when a field F_2, somewhat larger than the threshold field F_t, is applied to the specimen (Fig. 48). Let the field first be again distributed along the specimen uniformly. Let there again be a space charge on the borders of some region (the fluctuation of the field). The field within this region will be stronger than outside it (Fig. 49). But now, for $F > F_t$ a *smaller drift velocity will correspond to the larger field!* The fluctuation moves more slowly than the electrons both on the left and on the right of it.

As a result of it the excessive electrons of the fluctuation will be joined by new electrons coming from the left. On the right, where there was a deficit of electrons, that deficit will increase: the electrons "running away" from the slowly moving fluctuation. So the space charge on the boundaries of the fluctuation will increase. The field within the fluctuation will grow. But with $F > F_t$ the decrease of the drift velocity corresponds to the growth of the field, as a result of which the fluctuation will move still more slowly. There will be still more electrons stored on the left of the fluctuation and they will be still more scarce on the right. The field within the fluctuation will become still stronger and so on.

So when the drift velocity of electrons decreases with the growth of the field, the uniform distribution of the field proves to be unstable with respect to the small fluctuation of the field: having appeared, this fluctuation begins growing. When will the growth of the space charge on the fluctuation boundaries stop? And, consequently, when will the fluctuation growth cease? That will happen when the velocity of electrons within the fluctuation equals the velocity of electrons outside the fluctuation.

Until the excessive field within the fluctuation is small, the voltage drop on it is also very small, and its presence in the sample does not actually affect the velocity of electrons outside the fluctuation (Fig. 49). But as the field within the fluctuation is growing, the voltage drop on the fluctuation is always larger and larger. Since the bias voltage applied to the sample is constant, the share of the voltage received by the rest of the specimen is much less. The field outside the "fluctuation" is falling. We have written fluctuation in inverted commas in the previous sentence. Indeed, the word fluctuation means " a slight random deviation", and it is somehow inconvenient to use it when denoting something which consumes the greater part of the voltage applied to the sample. Such a large fluctuation is in fact a domain. Unlike the situation within the domain, the velocity of electrons outside the domain falls with the decrease of the field, because the field outside the domain is always less than F_t. So, when the voltage drop on the domain is large, the velocity of electrons outside the domain decreases. When the velocity of electrons outside the domain becomes equal to the velocity of the domain, the growth of the latter comes to an end.

The stable region of a strong field, "running" in the specimen from the cathode to the anode is nothing else but a domain, which was observed by Gunn in his experiments.

The maximum field in the Gunn domain F_m can be very large. With an average field in the specimen $F_t = V_t/L \cong 3.2$ kV/cm, the value F_m can make 20–300 kV/cm.

Applications of Gunn Effect

Many years have passed since Gunn discovered the current oscillations in GaAs and in InP. Thousands of articles, devoted to the investigation of the Gunn effect and to various physical phenomena connected with it, have been published in numerous scientific journals.

By now the Gunn effect has been observed in more than twenty semiconductor compounds: indium antimonide (InSb), indium arsenide (InAs), in germanium subjected to compression, in silicon cooled to low temperatures, in semiconductor compounds: GaAlAs, InGaSb, InAsP, InGaAsP and in many other semiconductors. In each of those semiconductors the researchers try to calculate theoretically and to measure experimentally the dependence of the drift velocity of electrons upon the electric field $v(F)$. (As we have seen, this dependence plays a key role in the Gunn effect.) Knowing $v(F)$, it is possible to calculate the dependence of the domain parameters: the field in the domain, the size and velocity of the domain on the length of the specimen, the concentration of electrons in it and the applied voltage.

The domain of a strong field, moving very fast, can be used to modulate beams of light, to generate rather powerful high frequency acoustic fluxes (ultrasound), to code and decode information in communication systems and for many other purposes.

But the main application of the Gunn diodes is the microwave generators.

When the radar pulse reflected from the target returns to the radar aerial installment, it is necessary to amplify it. A sensitive modern amplifier comprises a comparatively low-powered generator, very stable and of low noise (a heterodyne). The Gunn diodes are widely used as heterodynes in many modern radars (Fig. 50).

Thousands of people perish every year in car accidents. One of the main reasons for these accidents is exceeding the speed limit. In towns there are traffic control systems, supplied with special radars which register every case of speeding. The Gunn diodes are often used in such radars as generators. The Gunn diodes have been used in a special traffic system called "the pseudo-policeman". Thousands of small metallic boxes are placed along the highway, near enough to see one box from the other. One of a few scores of boxes is equipped with an automatic radar. If a car exceeds the speed limit within the visibility of the radar , the camera registers the license number of the car, and later on the driver is handed a summons obliging him to pay a fine.

The driver does not know which of the boxes is equipped with a radar (the police regularly change the position of the radar, moving it from box to box). The driver therefore takes care not to exceed the speed limit throughout the way.

Fig. 50. Applications of the Gunn diodes. The portable system of the radio-telephone communication between the frontline positions and the battalion headquarters. The Gunn diode is used as an amplifier and generator (*Electronics*, 1969).

CONCLUSION

One of the largest and most stormy oceans in the world of science is the Ocean of Physics. The Sea of Solid-State Physics, which is part of this

Ocean, has been investigated and studied rather thoroughly. But let us not be deluded. According to statistics, an impartial and exact science, the greatest number of shipwrecks takes place neither in the waters of the mysterious Bermuda triangle, nor near the bleak shores of the Antarctic. The majority of shipwrecks take place close to the densely populated and well-cultivated coasts of the Channel. The Sea of Solid-State Physics is rich in undercurrents, in reefs and shoals, and very often in unforeseen depths.

We have visited one of the bays of the Sea of Solid-State Physics – the Bay of Semiconductors. Can we say that we have acquired a good knowledge of it?

We have covered the most popular tourist routes along the coasts of the bay. We have flown round the bay by helicopter. We also made a few launch cruises on a fine sunny day.

For the first time that might be enough. But if you have enjoyed it, don't forget that you have just caught a glimpse of a vast and curious part of the bay. We have not yet reached one of the most important and interesting regions – that of Diodes and Transistors. A high Barrier of *p-n* Junction prevented us from reaching it. Another Barrier – the Metal-Semiconductor Barrier barred our way to the Schottky Diodes, to Field Transistors, and to some other very important devices.

We have not yet mentioned some other beautiful and mysterious realms. The physics of semiconductors has become a science so extensive, its applications so wide and varied, that a lifetime might not be long enough to form a close acquaintance of the Bay of Semiconductors.

And besides, there are many other bays, seas and oceans also rich in undercurrents, in sharp coral reefs and beautiful beaches. They might also attract you.

But one should remember that the Ocean of Science is vast and universal, and the bottle thrown in the Bay of Semiconductors may be found in the Oceans of Biology or Chemistry, or even in the Ocean of Social Science. And the Bay of Semiconductors is visited by very large fish and sea animals from other Seas and distant Oceans.

Whatever sea you might choose – happy sailing!

May success be yours!